SOIL RESOURCES
AND THE ENVIRONMENT

Soil Resources and the Environment

U. ASWATHANARAYANA

Adviser on Environment & Technology
c/o Ministry of Environment, Maputo, Mozambique

(Former Commonwealth Visiting Professor,
Universidade Eduardo Mondlane, Maputo, Mozambique)

Science Publishers, Inc.
U.S.A.

SCIENCE PUBLISHERS, Inc.
Post Office Box 699
Enfield, New Hampshire 03748
United States of America

Internet site: *http://www.scipub.net*

sales@scipub.net (marketing department)
editor@scipub.net (editorial department)
info@scipub.net (for all other enquiries)

© 1999, U. Aswathanarayana

Library of Congress Cataloging-in-Publication Data

Aswathanarayana, U.
    Soil resources and the environment / U. Aswathanarayana.
      p.   cm.
    Includes bibliographical references (p.268).
    ISBN 1-57808-067-3
    1. Soil science.   2. Soil management.   3. Soil ecology.
  I. Title.
  S591.A85   1999
  631.7--dc21   99-39258
                                                           CIP

ISBN 1-57808-067-3

Published by Science Publishers Inc., Enfield, NH, USA
Printed in India.

To Vijayalakshmi, Mother of my children,
Viswanath, Vasundhara, Srinivas, Indira and Vani,
who together brought me much happiness.

Preface

The tenet of the great Chinese philosopher, Confucius (551 - 479 B.C.), that "the essence of knowledge is, having it, to apply it" appears simple *prima facie,* but has the following profound implications built into it : (i) that acquisition of knowledge may not be regarded as an end by itself, but as a means to an end, (ii) that knowledge acquired is meant to be applied, and (iii) that the person who possesses knowledge is under moral obligation to apply such a knowledge for some practical purpose. In the context of this kind of approach, knowledge *ipso facto* refers to knowledge of the physical world, since metaphysical knowledge is not meant to have earthly (literally also, in the present case) purposes.

Sustainable Development is meant to be directed towards " improving the quality of human life while living within the carrying capacity of the supporting ecosystems". The present volume is guided by this perspective, when it looks at soil from two interrelated angles, namely, as a resource and as an environmental medium.

Soil has multiple uses. Most important of all, it is necessary for plant growth. Soil is a source of the sediment. It acts as a filter for the groundwater. Soil has been used to construct human shelters from time immemorial. Stabilized soil blocks are currently used for the construction of walls, houses, earth dams, and water tanks. Soil serves as a bearing material for roads, pipelines, and buildings. It is also a direct source of some ores, such as bauxite (Al-ore), garnierite (Ni-ore), and lateritoid gold.

Whatever the technological advances achieved, soil will always be necessary for humans to grow most of the food, fodder and fiber they need. About 99 % of our food comes from land and grains provide at least 80 % of the food worldwide. The food requirements of the projected population of 10 billion in 2050 is estimated to be 3.5 b.t. per year. This is at once a profound challenge and a great opportunity. It is only through an understanding of the complex processes that go on in the soil, and the interactions of the soil with water, air and biota, can we hope to optimize the biological productivity of the soil, while ensuring ecological sustainability and protecting the soil environment.

A worldwide consensus is emerging in favor of teaching Geology as an earth system science linked to such cognate subjects as Agriculture, Land-use and Meteorology, and focused on the ecologically sustainable management of water, soil, and mineral resources. This book is a humble attempt to provide a teaching tool for the proposed mode of instruction with regard to soil resources.

The present volume is characterized by a broad-spectrum, contextual approach. It emphasizes the dynamics of the soil processes and is modeling oriented. It seeks to provide an overview as to how knowledge of the soil processes can be put to practical use to improve agricultural productivity, reduce soil degradation, build low-cost habitations, conserve water, and prevent contamination of the soil. Attention is drawn to the impact of climate change on the soil processes and productivity of ecosystems, and ways and means of mitigating its adverse consequences. The present volume can be used as an undergraduate textbook/reference book in earth sciences, geography, ecology, pedology, agriculture, and civil engineering. It may be useful to resource managers and administrators who are concerned with soil as a resource or as an environmental medium.

In keeping with its interdisciplinary character the book carries two Forewords by savants in the areas of agricultural science and geochemistry respectively. Prof. M.S. Swaminathan, Unesco - Cousteau Professor of Ecotechnology, is my ideal scientist - brilliantly original while being practical, self-effacing, always-smiling and ever-helpful. Posterity will long remember with gratitude his great contribution to the Green Revolution in India and other developing countries. Prof. Gunter Faure of Ohio University, is a teacher, researcher, editor, author and administrator *par excellence* in the area of geochemistry. Under his dynamic leadership the International Association of Geochemistry and Cosmochemistry, is promoting globally the dissemination and application of geochemical knowledge. Professors Jul Låg (Oslo,Norway), and R.Vaidyanadhan (Cuddalore,India), made useful suggestions. I am particularly beholden to Dr. Maurits van den Berg (Maputo, Mozambique) who reviewed the manuscript and provided literary sources, and exercise material. The meticulous editing of the manuscript by Margaret Majithia undoubtedly improved the consistency of presentation.

The author was sustained by the small but selective collection of books in science and technology in the British Council library in Maputo, before the Authorities in London virtually closed down the only decent library in Mozambique on grounds of economy.

A good turn is always context-specific. In the context of Mozambique, the greatest good turn is assistance to survive! Mahomed Essak of Mozambique Services (MS), Luis de Macedo of Louis Berger International Inc., and Francisco Mabjaia of the Environment Ministry, made it possible for me to continue in Maputo and complete the book.

"The traveller has no destination; the one which seemed so, turned out to be another road "—Urdu couplet by Abdul Halim Jaffar Khan.

Maputo, Mozambique   U. Aswathanarayana
Oct. 1998

Foreword

Since World War II nearly one billion ha. of good farm land has been degraded due to a wide variety of human-induced causes. Quite often unsustainable pathways of development and irrigation programmes have led to the spread of soil health problems such as salinity, alkalinity and waterlogging. It is now widely recognized that the future of global food security depends upon our ability to manage our soil and water resources based on principles of ecology, economics and equity. The present publication of Prof. Aswathanarayana is therefore a timely contribution.

As mentioned in the Preface of the book, the essence of knowledge is "having it, to apply it". The book should help all those interested in soil health care and soil resources management to achieve their goals. In particular, this is a valuable book for introducing low external input sustainable agricultural techniques. We should now initiate a dynamic program to conserve the fertility of prime farmland and restore the production potential of degraded soils. This is one of the greatest challenges for sustainable food security in the coming millennium. A global soil health care programe would need for its success a clear understanding of the physics, chemistry, and microbiology of soils. This book provides us with detailed information on the geochemical aspects of soil resilience and fertility management.

We owe Prof. Aswathanarayana a deep debt of gratitude for this labor of love.

Madras, India   **M.S. Swaminathan**
                    UNESCO-Cousteau Professor of Ecotechnology

Prologue

As we conclude the twentieth century, we face serious problems that arise from the rapid growth of the human population compounded by the legitimate expectations for a better standard of living, including : better nutrition, better medical care, and better housing. The rising expectations of the world's growing population collide with several physical conditions that characterize our existence on the Earth: 1) Metallic and non-metallic mineral deposits are finite and non-renewable; 2) Fossil fuels will approach exhaustion in the next century; 3) Waste products generated by the growing human population are contaminating continents and oceans; 4) the rate of food production is lagging behind the growth of the population; 5) Global warming of the world's climate will cause flooding of living space and may require a redistribution of the human population.

A linear extrapolation of the present trends into the future leads to predictions of large-scale starvation, wide-spread poverty, and the resulting conflict between wealthy and underprivileged nations. Fortunately, the potential problems outlined above have been recognized by the leaders of the world and there is reason to believe that a catastrophe can be averted. In fact, the recognition that all of us live in the same house has motivated people to take certain actions which collectively add up to major changes in the way we live. These actions include : 1) Recycling of metals, plastics and paper; 2) Efforts to avoid contamination of water; 3) Conservation of energy; 4) Cleanup of the contaminated sites to their original condition; and 5) Research to develop new sources of energy and to increase food production. These and other measures may deflect us from the road to disaster, especially if the human population stabilizes as a consequence of voluntary choices made by individual human beings.

These considerations are the background for the present book by Professor Aswathanarayana in which he addresses the importance of soil conservation to the growth of food crops around the world. In the best traditions of science in the service to humanity, Professor Aswathanarayana points out how various anthropogenic practices affect the fertility of the soils. Some of these practices are beneficial in the short term, but have undesirable long-range consequences that should be avoided. At a time when food production needs to rise to meet the growing demand, we must adopt the most appropriate farming methods and take care not to allow soils to be degraded by shortsighted and uninformed practices.

The challenge we all face is to reach a state of stability in the population of the world and hence to achieve a world economy that can sustain that population for all time. This book is a valuable contribution towards this goal.

Columbus, Ohio, USA   **Gunter Faure**
March 30, 1999   Professor
                                          The Ohio State University

Acknowledgment

Grateful acknowledgment is made to the publishers, authors/editors of the journals/books from which some figures that appear in this volume have been redrawn or adapted. The figures in the volume so drawn are shown in parentheses after the publisher of the concerned book or journal and the year of publication.

A.I.M.E. - *Int. Laterite Symp.*, 1979 (10.6, 10.7)
Applied Publishers - *Introduction to Exploration Geochemistry*, 1974(1.3)
Colorado School of Mines - *The Application of Geochemical Techniques to Mineral Exploration*, 1968 (2.1)
CRC Press - *Geomedicine*, 1990 (9.1)
Estuaries - v.18, pp. 648-659, 1995 (5.1)
FAO - Report no. FHE/72/9, 1972 (3.9)
George Allen and Unwin - *Environmental Chemistry*, 1985 (3.2, 3.8)
ILO - *Small-scale Manufacture of Stabilized Soil Blocks*, 1987(7.3, 7.5)
IPCC - *Climate Change* 1995, 1996 (5.2)
Longmans - *An Introduction to Soil Science, 1993,ELBS 2nd ed.* (1.1,2.3)
Water Resources and Agricultural Development in the Tropics, 1987 (2.2, 6.4, 6.6, 6.9)
McGraw-Hill - *Principles of Chemical Sedimentology*, 1971 (10.2)
Nat. Acad. Press (USA) - *Saline Agriculture*, 1990 (5.4)
Oxford Univ. Press - *The Chemistry of Soils*, 1989 (2.4, 2.5, 3.5, 3.7)
Pentech Press - *Soil Laboratory Testing*, v.1, 2nd ed., 1992 (7.1,7.2, 7.4)
Pergamon Press - *The Heavy Elements : Chemistry, Environmental Impact and Health Effects*, 1990 (2.6, 2.7)
Routledge - *Mountain Environments in Changing Climate*, 1994 (5.3)
Science - v. 189, pp.550-551, 1975 ,(3.1)
SKAT - *Appropriate Building Materials*, 1993 (7.6)
Soil.Soc. Amer. J., v.47, pp.217-220, 1983 (3.6)

Springer-Verlag - *The Importance of Chemical Speciation in Environmental Processes*, 1986 (8.1); *Speciation of Metals in Waters, Sediments and Soil Systems*, 1987 (8.2)

UNEP - UNESCO - *Man's Dependence on Earth*, 1987 (1.5, 1.6)

UNESCO - *A Global Geochemical Database - for Environmental and Resource management*, 1995 (1.4, 6.3)

Univ. of Linköping - Tema V, Rept. 14, 1990(6.5)

World Bank - *The Efficient Use of Water in Irrigation*, 1987 (4.1, 4.2, 6.1, 6.2, 6.7, 6.8)

Contents

CHAPTER

1

Introduction

To understand a problem, is to partially solve it
—Chairman Mao of China

1.1 SOILS AND MAN

The progression of hominids to modern *homo sapiens* during the last four million years is marked by some profound "benchmarks", namely, Upright walker, Tool maker, Fire builder, Cave painter, and Stargazer. The basic requirement to qualify for inclusion in human lineage is standing upright and walking on two feet. Paleoanthropologists have given the following dates for the "benchmarks"

2.6 m.y.   — earliest known stone tools,
1.0 m.y.   — earliest evidence of controlled fire,
340,000 y — earliest evidence of shelters,
150,000 y — anatomically modern *homo sapiens*,
125,000 y — earliest evidence of modern humans in Africa,
100,000 y — earliest known burials,
30,000 y   — earliest known cave paintings,
10,000 y   — end of the most recent ice age,
5000 y   — first cities.

(*source : National Geographic*, USA, Feb. 97).

In terms of civilization, *Homo sapiens* who survived solely by hunting and food gathering, are indistinguishable from other animals. Modern civilization can be deemed to have dawned about 10,000 years ago, when man learned to cultivate land to grow food. It is no accident that the fertile alluvial valleys of the Nile, Euphrates-Tigris and Indus constituted the early sites for significant human settlements and civilizations. For instance, the earliest village at Marimda in the Nile Delta is said to be

7,000 years old, i.e., 5000 B.C. The earliest irrigation ditches were dug there around 3000 B.C. (presently, there are 16,000 km of irrigation canals in the Nile Delta).

The need to measure the dimensions of land under cultivation, led to the development of geometry. Domestication of animals and building of canals made for a more productive agriculture. Larger populations could then be supported because of the availability of greater quantities of food. The use of human and animal muscle power, and burning of animal dung and agricultural refuse as fuel, resulted in the rise of energy consumption.

Soil is a convenient material for building shelters. It is ubiquitous and more convenient to handle than stones. Man must have started building mud houses at about the same time as he started cultivation of crops. Mud-built houses do not survive for more than a few years, which explains why we do not find relics of mud-built walls. Burnt bricks made from soil survive much better, as is evidenced by the excellent preservation of brick structures built by the sophisticated urban societies of the Indus Valley around 2500 B.C. There is archeological evidence in India to show that some mud-built forts in which gluey compounds (such as animal dung, ant heap material, animal blood) were used to stabilize the soil, survived for a few hundred years. Mud-built houses are a common sight in rural areas of the developing countries. In parts of West Africa (e.g. Mali), elaborate and esthetically elegant mud-built structures continue to be used for habitation, business premises, mosques, etc.

Copper and iron occur in the native form in the soil. It is hence not an accident that the early civilizations used copper and iron picked up from the soil, to make implements, weapons, and ornaments. Gold also occurs as nuggets in the soil, but they are very rare. There is at least one instance (Yellowknife region in the Northwest Territories in the Canadian Arctic), where the local Eskimos used to fashion their knives from gold nuggets, as gold was the only metal available!

In the Hindu culture, earth (i.e. land/soil) is worshipped as a goddess. In one of His manifestations, Lord Vishnu has two wives: temperamental Sridevi (goddess of wealth) who will withdraw her favors if a devotee is not sufficiently worshipful, and the placid Bhudevi (goddess of earth) who patiently bears the filth of mankind, just as a mother would in the case of her dirty child. Bhudevi is worshipped at the start of the agricultural season, and while laying foundations for new structures.

Romans worshipped Sterculius as the god of soil fertility. Ceres is the Roman goddess of grains (hence, the word cereal for wheat, rice and maize).

In the ultimate analysis, man derives all the nutrient elements from the soil. However, he does not get them by eating the soil, but by eating the plant that grows in the soil, or by eating the meat of the animal that eats the plant (geophagy, or eating of soil as food supplement, is practiced in some cultures, by some categories of people, such as pregnant women, but this is hardly the norm). Soil is both a resource which has to be used wisely to grow food and also an environmental medium which has to be monitored to prevent the adverse impact of man. The adverse impact may manifest itself in the form of soil erosion, exhaustion of soil fertility, salinization and desertification, and the pollution of soil. Polluted soil contaminates the water and biota, and the pollutant ends up in the food chain. It should not be forgotten, however, that man has had a beneficial impact on the soil also, through drainage, fertilization and irrigation, which have improved many soils initially not capable of supporting the growth of crops. Thus, an ecologically sound management of the soil has two objectives: to support sustainable agriculture and to prevent soil pollution.

1.2   SOIL FORMATION FACTORS

Soils are the loose mantle on the surface of the earth. Chemically, soils may be defined as multicomponent, open, biogeochemical systems. Because they are open systems, soils continually exchange both matter and energy with the surrounding atmosphere, hydrosphere, and biosphere.

Soil formation is just the last stage of the weathering process. Disintegration and decomposition of solid rocks lead to the formation of a blanket of unconsolidated material on the surface of the earth. This is called the *regolith*. Soil constitutes the upper part of the regolith and is composed of not only mineral matter, but also organic matter, water, and air. To put it simply, if the products of weathering remain *in situ* above the weathered parent rock, they constitute the soil; if such products are transported by water, wind, etc. and deposited in water in a more or less layered form, they are called sediments.

The processes of soil formation are schematically shown in Fig. 1.1.

The nature and rate of soil formation depend upon the integrated effects of climate, nature of the parent rock, biological activity, topography, and time. Dokuchaev, the great Russian pedologist, demonstrated that soil formation is strongly related to climate (in terms of temperature, rainfall, wind, and potential evapotranspiration). Thus, the same parent rock may give rise to different types of soils, depending on the climate.

Climate

The influence of climate on soil formation arises out of the following considerations : (i) Temperature has a direct effect on the kinetics of the

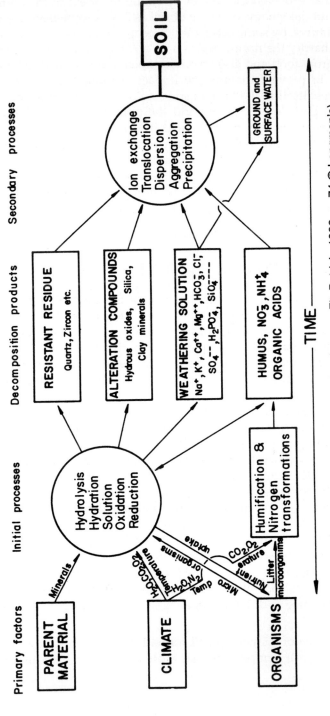

Fig. 1.1 Processes of soil formation (*source:* FitzPatrick, 1993, p. 74 © Longman's).

chemical reactions related to soil formation; (ii) Most chemical reactions involved in soil formation occur in the soil-liquid interface. If there is no moisture (originating from precipitation), it is not possible for weathering to take place; (iii) Precipitation and evaporation dictate the moisture dynamics in the soil formation, i.e., whether the soluble weathering products are removed by percolating (or runoff) water, or whether they accumulate in the soil; and (iv) Biological activity, an important factor of soil formation, is strongly influenced by climate.

The rates of soil formation are obviously related to rates of weathering. According to Strakhov, the rates of weathering in humid tropics may be 7–14 times, occasionally even 40 times higher than those in cool, temperate climates.

Parameter	Humid tropics	Cool, temperate climate
Temperature (°C)	24 – 26	< 10
Rainfall (mm)	1,200 – 3,000	300 – 700
Organic matter production (t ha^{-1} y^{-1})	ca. 100	8 –10

Even if one does not fully accept Strakhov's figures, there is little doubt that weathering and soil formation are far more intense in the humid tropics than in the cool, temperate climates. For instance, tropical climate is responsible for the formation of about 40 cm of soil in 45 years since the Krakatoa volcanic eruption in Indonesia, whereas hardly any soil has developed during the last 10,000 years since the last glaciation in the Canadian Arctic.

Biological activity

Biological processes involving plants and microorganisms play a crucial role in the formation of soils. Plants are the main source of organic matter in the soil. Mesofauna, e.g. worms, ants, and termites, play an important role in soil homogenization and structure stabilization. That physical and chemical processes alone (i.e., unaided by biological processes) cannot lead to the development of soils, can be seen from the absence of soils in hot or cold deserts. Deep-rooted trees are capable of removing significant quantities of dissolved metals (through base exchange around rootlets) from depths of as much as 30 m. Part of this matter is returned to the surface as leaf fall. Humic and fulvic acids are capable of removing metal cations by chelation. Figure 1.2 shows (i) how the presence of organic acids enhances the solubility of Al and Fe ions, and (ii) how the higher solubility of Al vis-à-vis Fe at the same level of organic acid content (16 ppm C) and pH (7 to 9) can bring about the separation of Al from Fe.

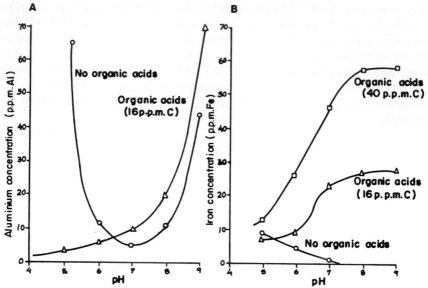

Fig. 1.2 Solubility of aluminum and ferric iron as a function of pH and the presence of organic acids (*source:* Ong et al., 1970)

Microorganisms are capable of augmenting the processes of inorganic oxidation and reduction. For instance, *Ferrobacillus* bacteria can oxidize ferrous iron, and *Thiobacillus* can oxidize sulfides to sulfates. On the other hand, anaerobic bacteria, such as *Desulphovibrio* and *Desulphotemaculum*, can reduce sulfate to sulfide.

Parent material

The nature of the parent material also has a bearing on the nature and rate of soil formation. Let us take a quartzose sandstone and a basalt in a given climatic situation (say, humid tropics). The quartzose sandstone can yield only quartz on disintegration. The resultant soil does not permit plant growth to any significant extent, and thus severely limits the role of organic matter in the soil formation. On the other hand, the weathering of a basalt leads to the production of clays which support luxuriant vegetation, with its attendant consequences in accelerating the formation of the soil. Soils generated from limestones tend to be alkaline, but the acidity produced by decaying organic matter tends to neutralize it in part. Other things being equal, soil develops more readily on a porous, soft or fractured rock, than on a hard, massive rock.

Topography

According to Loughnan, the gradient of slope and the elevation determine (i) the extent of surface runoff and the percolation of rain water and hence the amount of moisture retained by the rock, (ii) the rate of leaching of soluble constituents by the moving soil and groundwater, and (iii) the rate of exposure of fresh rock arising out of the removal of weathered products. Hence, steep, stony situations have hardly any soil cover, whereas slumping and valley fill tend to permit the development of soil.

Time

Time is an important factor in soil development. Other things being equal, the longer the period of soil development, the more *mature* the soil will be.

1.3 SOIL PROFILE

A *soil profile* in the case of mineral soils comes into existence when weathered debris is transposed from one level to another, leading to the production of layers, called *horizons*. Some broad horizons — O, A, B, C — are generally recognizable above the bedrock or parent material (horizon, R). These are subdivided into a number of subhorizons, which have no universally agreed nomenclature. The present discussion follows the nomenclature adopted by FAO (1991). A description of the subhorizons and the geochemical processes leading to their formation are given in Table 1.1. The soil horizons may actually be more complicated than indicated. Additional horizons may exist. Some horizons may be poorly developed or absent altogether. Complications may arise by the changes in the relative dominance of physical and chemical weathering during the time the soil was forming. The thickness of the individual horizons may vary depending on the climate, nature of the bedrock, biological activity, duration of development of the soil, etc.

An example of the mineralogical and chemical characteristics of the different soil horizons are given in Fig. 1.3. The following variations may be noted top-down: (i) pH increases in the B horizon and remains virtually unchanged from then onwards, (ii) A_0 horizon has the highest content of organic matter, (iii) The clay content and the content of base metals reach their peak in the B_1 horizon, (iv) The cation exchange capacity reaches maxima in A_0 and B_1 horizons, and (v) The percentage of sand and clay is minimum in the B_1 horizon.

1.4 SOIL CLASSIFICATION

Communication with respect to soils has been hampered consequent to some countries, notably the USA, France, U.K., and Brazil, adhering to

Table 1.1 Geochemical processes involved in the formation of different soil horizons (FAO, 1991, nomenclature is used for soil horizons)

Horizon/ layer	Description	Geochemical Processes involved
O	Layer dominated by organic material, consisting of undecomposed or partially decomposed litter (such as leaves, needles, twigs, moss and lichens) which has accumulated on the surface, where it is not saturated with water for prolonged periods.	Thick O layers develop when the decomposition rate of organic matter decreases due to adverse conditions for biological acitivity, such as soil acidity, low soil fertility, and low temperatures. The O horizon is generally lost in a few years when the soil is used for agriculture.
H	Layer dominated by organic material, formed due to accumulations of undecomposed or partially decomposed organic material at the soil surface, where it is saturated with water for prolonged periods.	H layers form due to reduced decomposition rate of organic matter arising from adverse conditions for biologic activity due to anaerobic conditions for prolonged periods . On drainage, the decomposition gets accelerated, and H horizon is gradually lost.
A	Mineral horizon, formed at the surface, or below an O horizon, characterized by a high content of humified organic matter, intimately mixed with mineral fraction.	Horizon of maximum biological activity, resulting in mineralization and humification of organic matter.
E	Mineral horizon in which the main feature is loss of silicate, clay, iron or some combination of these, leaving a concentration of sand and silt particles.	Low pH, permanently humid conditions, and low temperatures, favor formation of this horizon. These conditions promote formation of humic acids in the overlying A horizon or O layer. The mineral matter gets partially solubilized, and moves downward through percolating waters.
B	Horizon formed below an A, E or O horizon, without rock structure, characterized by either (i) illuvial concentration of silicate clay (Bt), sesquioxides (Bs), humus (Bh), etc. or (ii) residual concentration of sesquioxides (Bo), or (iii) alteration that	Illuvial concentrations arise by precipitation of sesquioxides, and humus from the overlying E horizon. Residual concentrations of sesquioxides are common in tropical, humid conditions, when intensive weathering of primary minerals and leaching of liberated silicates, leaves behind resi-

(Contd.)

Table 1.1 (*Contd.*)

Horizon/ layer	Description	Geochemical Processes involved
	results in the formation of silicate clay (Bt) or liberation of oxides or both, and that forms granular, blocky, or prismatic structure (Bw).	dues of insoluble Fe and Al oxi-hydroxides. When the leaching is less intensive, weathering of primary minerals is accompanied by *in situ* formation of clay minerals. Structure formation is primarily a consequence of shrinking and swelling of clay minerals under alternating humid and dry conditions.
C	Horizon or layer consisting of unconsolidated material, that is little or not affected by pedogenic processes. Secondary accumulation of carbonates (Ck), gypsum (Cy), or soluble salts (Cz) may occur.	C horizons consist of loose, partly decayed rock which maintains its original structure, or the structure of unaltered, or unconsolidated parent material (in the case of sediments). Presence of primary accumulations of salt, gypsum or carbonates reflects the composition of the original material. Secondary accumulations occur in (semi-) arid regions, especially when the groundwater levels are high.
R	Hard bedrock, underlying the soil.	Unweathered bedrock or parent material.

their own national systems of soil classification. The Legend of the Soil Map of the World (FAO-Unesco, 1974), developed for a global soil map on 1 : 5 million scale, and revised in 1988 (FAO-Unesco, ISRIC 1988), is widely used as an international soil classification system. It consists of 28 Major Soil Groupings. Each Major Soil Grouping consists of several Soil Units. The soils are classified according to their diagnostic horizons and diagnostic properties, which are morphometrically defined, i.e., in terms of objectively observable or measurable characteristics, correlatable to their mode of formation. Driessen and Dudal (1989) lumped the Major Soil Groupings into 9 larger groups, based on the major factors that conditioned their formation. A summary of these groups and their components is given in Table 1.2.

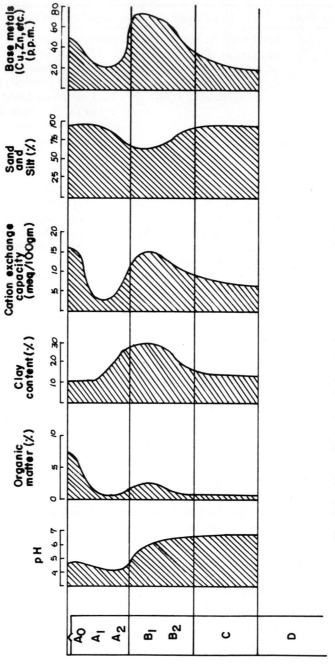

Fig. 1.3   Mineralogical and chemical characteristics of different soil horizons (*source:* Levinson, 1974, p. 98; © Applied Publ. Ltd.).

Table 1.2 Major Soil Groupings of the FAO — UNESCO System (1988) (source: Driessen and Dudal, 1989). The number in parentheses against each soil grouping is the percentage distribution of that soil in the world (*source:* FAO, 1991)

Major soil groupings	Description
1. **Organic** soils — *Histosols* (2.2)	Peat and muck soils; derived from incompletely decomposed plant remains, with or without admixtures of sand, silt or clay; majority of Histosols formed in boreal regions.
2. Mineral soils in which soil formation is conditioned by **human influences** — *Anthrosols*	Derived from various parent materials, modified by man through deep cultivation or through addition of material from elsewhere.
3. Mineral soils in which soil formation is conditioned by **parent material:** in volcanic material — *Andosols* (0.9), in residual and shifting sands — *Arenosols* (7.1) in expanding clays (2:1 lattice clays, smectites) — *Vertisols* (2.7)	Andosols are derived from pyroclastic materials. They occur in the volcanic regions. Arenosols are formed upon weathering of old, quartz-rich soil material or rock under tropical conditions, or from shifting sands in deserts and beaches. Vertisols develop in (former) sedimentary lowlands, and denudation plains.
4. Mineral soils in which soil formation is conditioned by **topography/physiography:** in lowlands (wetlands) with level topography — *Fluvisols* (2.8), *Gleysols* (5.7) in elevated regions with nonlevel topography — *Leptosols* (13.1), *Regosols* (4.6)	Fluvisols are developed in alluvial deposits. Gleysols are wet, and form in depressed areas with shallow groundwater. In tropical and subtropical regions, they are widely used to grow rice. Leptosols are shallow soils that form in high-altitude areas with high rate of erosion. Regosols are finegrained, unconsolidated weathering material.
5. Mineral soils in which soil formation is conditioned by their **limited age:** *Cambisols* (12.5)	Cambisols are derived from a wide range of rocks in colluvial, alluvial or eolian deposits. Widespread occurrence. Characterized by the absence of appreciable quantities of illuviated clay, organic matter, aluminum and/or iron compounds.
6. Mineral soils in which soil formation is conditioned by **climate (and vegetation)** of wet tropical and subtropical regions: *Plinthosols* (0.5), *Ferralsols*,	Plinthosols are soils with clayey materials which become hard when exposed to air. Ferralsols are red and yellow tropical soils with high content of

(Contd.)

Table 1.2 (*Contd.*)

Major soil groupings	Description
(5.9), *Nitisols* (1.6), *Acrisols & Alsols* (7.9), *Lixisols* (3.5)	sesquioxides. Nitisols are deep red tropical soils with clay illuviation and shiny peds. Acrisols are strongly weathered acid soils with low base saturation. Alsols are acid soils with high levels of "free" aluminum. Lixisols are strongly weathered soils in which clay is washed down from the surface soil to an accumulation horizon at some depth.
7. Mineral soils in which soil formation is conditioned by **climate (and vegetation)** of arid and semiarid regions: *Solonchaks* (1.5), *Solonetz* (1.1), *Gypsisols* (0.7), *Calcisols* (6.3)	Solonchaks are saline soils, most common in seasonally or permanently waterlogged areas in arid or semiarid regions. Solonetz are soils with high sodium saturation. Gypsisols and Calcisols are soils with high accumulation of calcium sulfate and calcium carbonate respectively.
8. Mineral soils in which soil formation is conditioned by **climate (and vegetation)** of steppes and steppic regions: *Kastanozems* (3.7), *Chernozems* (1.8), *Phaeozems*, (1.2), *Greyzems* (0.3)	Kastanozems are dark brown soils, rich in organic matter. A large part of them has developed in loess. Chernozems are black soils rich in organic matter. Phaeozems are dark soils rich in organic matter, derived from predominantly basic material. Greyzems are dark soils with grey tinge due to the presence of bleached quartz sand and silt.
9. Mineral soils in which soil formation is conditioned by **climate (and vegetation)** of subhumid forest and grassland regions: *Luvisols* (5.1), *Podzoluvisols* (2.5), *Planosols* (1.0), *Podzols* (3.9)	Luvisols are soils in which clay is washed down from the surface soil to an accumulation horizon at some depth. Podzoluvisols are soils having the bleached eluviation horizon "tonguing" into the clay accumulation horizon. Planosols are soils with an eluvial horizon abruptly over a dense subsoil, typically in seasonally waterlogged flatlands. Podzols are soils with a subsurface horizon that has been strongly bleached by aggressive organic acids.

1.5 SOIL AS A RESOURCE

Zimmerman's dictum, "Resources are not, they become", has profound technoeconomic implications. According to him, what constitutes a resource is governed by two considerations: (i) knowledge and technical means must exist to allow its extraction and utilization, and (ii) there must be a demand for materials or services produced. It is therefore perfectly possible that what was yesterday a nonresource, can become a resource today because advances in science and technology made it possible for that substance to be put to economic use. For instance, it is now possible to grow salt-tolerant crops of food, fodder and fiber in saline soils which were previously considered uncultivable. Cassava is one of the few food plants that can grow in poor soils, under conditions of extreme drought. But in the process, it is rendered bitter and therefore unfit for human consumption. Technology is now available to use such cassava for the production of ethanol. In other words, technological advances have made it possible to make productive use of otherwise inhospitable soils, even under adverse weather conditions.

Any attempt to improve the productivity of the soil has to be based on an understanding of the processes taking place in the soil. Knowledge of the physical, electrochemical, biochemical and geochemical interactions taking place in the soil constitutes the basis for designing ways and means of mitigating soil acidity and soil alkalinity, improving bioavailability and the kinetics of the nutrient uptake by the plants in relation to soil fertility, environmentally sound practices of fertilizer and irrigation application, and sustainable agriculture packages (Låg, 1987; Sposito, 1989; Swaminathan, 1991; Wild, 1988, 1993; Aswathanarayana, 1995).

All land on earth is not suitable for agriculture. Land may be covered by ice, it may be too cold or too dry, the slope of the land may be too steep to permit agriculture, or the soil may be too shallow or too poor or too wet, or the land may be inaccessible. For these reasons, out of the total land area of 13.9×10^9 ha on earth, only 22% of it (about 3.3×10^9 ha) is suitable for agriculture. Technological advances are changing this situation, however. For instance, drip irrigation makes it possible to grow plants almost anywhere—in rugged terrains, in sandy soils of low moisture capacity, and in arid climates of high evaporativity, and so on.

There is much variation in the productivity of agricultural land. The area of agricultural land with a high level of productivity is only 100×10^6 ha (i. e., about 3% of total land area).

The land availability (ha/capita) and the percentage of potentially arable land in different continents is given in Table 1.3 (Buringh et al., 1975).

Soil Resources and the Environment

Table 1.3   Land availability in different continents

Continent	Land availability (ha/capita)	Percent of potentially arable land under cultivation
North America	1.84	43
South America	2.81	12
Africa	1.85	22
Europe	0.60	54
Asia	0.39	77
World	0.88	42

Thus, South America has considerable potentially arable land which could be cultivated, whereas most of the arable land in Asia is already under cultivation.

In the case of Africa, only 22% of the arable land is under cultivation, though in countries like Egypt, practically all the arable land is under cultivation (only 3% of the land in Egypt is arable). For instance, in the case of Mozambique, the problem is not the availability of arable land *per se* (about 2 ha of arable land per capita), but its extremely low productivity ($0.5 - 1$ t ha^{-1} of, say, maize).

Cereal crops account for 75% of the world food production (dry matter basis) and root and tuber crops, about 7%. The increase in food production during the last 20 years has been achieved through increasing the yields rather than bringing more land under cultivation. The possibilities in this direction are evident from the particulars given in Table 1.4 (after Cook, 1979).

Table 1.4   Present and potential yields of cereal crops (t ha^{-1})

	Wheat	Rice	Maize
Potential yield (experimental station)	12.0	14.0	13.0
Top country (average yield)	5.2	6.0	5.7
Developed countries	2.2	5.7	5.0
Developing countries	1.2	1.9	1.3

The very high yields in the experimental stations are evidently due to a high technical level in all aspects of soil, water, and crop management. At the other end of the spectrum, the low levels of yields in the developing countries is attributable to poor management of soil, water, and crops. This is because the average farmer in the developing countries cannot afford the use of better methods. Buringh et al. (1975) believe that through the application of various technological inputs and adoption of correct agronomic practices, it is indeed possible to raise the world food production tenfold.

1.6 GEOCHEMICAL DATABASES FOR SOIL RESOURCES MANAGEMENT

Fortunately there is no conflict in the utilization of a soil for agriculture and civil construction purposes. Soils rich in organic matter are preferred in agriculture, whereas they are unsuitable for construction purposes. Similarly, soils with high sand content are preferred for construction purposes, whereas such soils tend to be infertile and unsuitable for agriculture.

In the past, the soil data gathered in the course of soil surveys was mostly used to characterize individual soil or mapping units. FAO (Food and Agricultural Organization, Rome) in collaboration with ISRIC (International Soil Reference Information Center) has developed a computerized Soil Database package (SDB) which facilitates the interpretative use of soil data for land evaluation and land-use planning (FAO Report no. 64, 1989). SDB is a stand-alone program, i.e., it does not need a supporting database management system. It is compatible with dBASE III/IV and can be used in conjunction with GIS (Geographic Information System) to prepare automated land evaluation packages.

Sims and Heckendorn (1991) gave a detailed account of methods of soil analysis. Darnley et al. (1995) explain how the developing countries can make use of the geochemical databases for the management of natural resources, such as soils. Thanks to the availability of inexpensive PCs, the developing countries can now afford to build and manage adequate geochemical databases to determine which soil is best suited for which purpose, and how to make optimal use of the soil. In the case of soils, the media that are sampled are: Humus, Upper A_{25} (residual and/or overbank/and/or floodplain) and Lower (C) horizons. Solid samples of soil, plus humus and water are analyzed by a suitable combination of the following methods: XRF, NAA, AAS, ICP-AES, ICP-MS, etc. National geochemical maps are produced on scales ranging from 1: 1 million, to 1: 50,000. The storage and management of large global or regional data sets have been rendered possible because of advances in high capacity optical storage systems. A variety of computer software is available for data analysis (DOS, Windows NT, UNIX, OS/2, System 7, VMS, etc.). Even though the data itself is in numeric form, the data fields are of the character type which can be transformed to numeric as and when needed.

The logical relationships between databases are given in Fig. 1.4. The recommended databases are: (i) an Index database, (ii) a Block database, and (iii) a Bibliographic database.

The *Index database* is meant to hold administrative (survey objective, geographic boundaries, characteristics of the region, etc.) and methodological information (sampling medium, type of sample, elements deter-

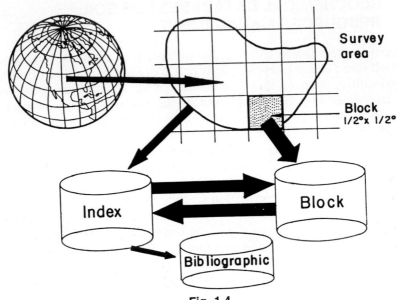

Fig. 1.4
(*source:* Darnley et al., 1995, p. 72)

mined, organization and contact person responsible, etc.) about national geochemical surveys. *The Block database* contains average elemental contents within $1/2° × 1/2°$ latitude-longitude blocks. The Northern and Southern hemispheres have been divided into 129,600 blocks each. The block area varies from 3,080 km^2 at the equator, to about 53 km^2 near the poles. Each block has been assigned a Morton number. The Block database contains the following information: the way a given block is linked to the parent survey, location of the block (Morton number), unique variable (element/anion), sampling media, size fraction, and analytical method (total, extractable, radiometric) employed, etc. The *Bibliographic database* contains references relevant to the surveys in the Index database.

1.7 SOILS AND LAND-USE

The terms "land" and "soil" are used more or less synonymously in this section.

The World Bank projects that between 1990 and 2030 the world's population will grow by 3.7 billion, demand for food will almost double, and industrial output and energy use will probably triple worldwide, and increase sixfold in the developing countries (source: World Development Report, 1992).

The United Nations Report entitled "Comprehensive Assessment of the Freshwater Resources of the World" (1997), draws a sombre picture of how water shortages in the twenty-first century could constrain economic and social development and become sources of conflict between countries. Water-use has increased twice as fast as population during this century. Pollution has made matters worse. Total water-use in the world rose from 1,000 $km^3 y^{-1}$ in 1940 to 4,130 $km^3 y^{-1}$ in 1990. It is expected to rise to 5000 $km^3 y^{-1}$ in 2000. At least one-fifth of the world's people lack access to safe drinking water. About 80 countries in the world, accounting for 40% of the world's population, are already suffering from serious water shortage, which has become a limiting factor in their economic and social development. The Report forecasts that by 2025 as much as two-thirds of the world's population will be affected by moderate to severe water scarcity unless appropriate mitigation measures are taken.

Presently, agricultural irrigation accounts for 70% of all water-use. Increasing population would need more food and hence more water would be needed for irrigation. The Report argues that water must be perceived as a marketable commodity, with its use being subject to the market laws of supply and demand. It has been estimated that in the USA, one acre-foot of water would yield an income of USD 400 when used in agriculture, and USD 40,000 when used in manufacturing. So, in the American context, when water is scarce, it would be used for drinking and manufacturing purposes. Food is to be grown in areas where water is plentiful.

A complicating factor is the global trend towards urbanization: urban population, as a percentage of total population, varies from 100% in city-states such as Singapore, 80 - 90% in several countries of Western Europe (e.g. U.K.: 89%) and South America (e.g. Argentina: 86%), 20 - 30 % in most African countries (e.g. Mozambique: 30 %), with Bhutan having the lowest percentage (5%). Globally, urban populations are increasing at a faster rate than the general population growth; this is particularly evident in the developing countries where this differential can be as high as 3 to 4. This has serious implications for land-use planning. Cities cannot be shifted to places where resources such as water and fertile soil are available. On the other hand, services such as potable water, habitation, sanitation, roads, etc. have to be organized keeping in mind the existing location of towns.

The high standard of living of Industrialized countries is based on the high rate of consumption of minerals and metals derived from rocks and soils. For instance, the annual per capita consumption of minerals in the USA is about 10 tons (crushed rock: 4,150 kg; sand and gravel: 3,890 kg; cement: 360 kg; clays: 220 kg; salt: 200 kg; phosphate rock: 140 kg; other nonmetals: 485 kg. — total 10,055 kg). The per capita consumption of metals is 610 kg (iron and steel: 550 kg; aluminum: 25 kg; copper: 10 kg;

lead: 6 kg; zinc: 5.5 kg; manganese: 6 kg; other metals: 8.5 kg) (Bates and Jackson, 1982). In the USA, about 6,600 L of water is used to grow one day's food for an adult.

Land is needed for growing crops and livestock, as well as for housing, industries, transport, recreation, waste disposal, and other services (Rabbinge and van Ittersum, 1994). The demands on land are increasing rapidly as increased populations seek ever higher standards of living. But the amount of land does not increase ("Buy land — they don't make it any more", Mark Twain advised a friend!).

There are situations wherein the nature of the soil may delimit its use more or less unequivocally. For instance, a soil rich in organic matter is suitable for agriculture, but is unsuitable for construction purposes, whereas a sandy soil which is unsuitable for agriculture because of its low fertility, is the preferred material for use in soil stabilization for engineering purposes. However, land-use conflicts will arise if different socioeconomic interests press for the use of a given piece of land, (say) to construct a housing colony, to grow crops and vegetables, or to serve as a waste dump, for all of which that land is suitable. Evidently, the same land cannot be used for the various purposes *all at the same time*. If a gravel pit is to be dug in a piece of land to provide gravel for constructional purposes, the same land cannot obviously be used for growing vegetables (a pit formed due to the removal of gravel can *subsequently* be used for recreational purposes, such as sunken gardens, and swimming ponds, but that is a different story).

In rural areas, agriculture would be the principal form of land-use, with most of the housing tending to be single-story. Consequently, the focus will be on the chemical and biological characteristics of the soils which determine the fertility of the soils and their suitability for growing different crops. This is, of course, an oversimplification. For instance, the bearing strength of the soils has to be taken into consideration if agriculture machinery with high axial loads is proposed to be used in farming.

Additionally, the following particulars about the soils need to be known in order to determine the appropriate land-use (Archer et al., 1987, pp. 191-192): soil profile, soil slope, cultivability, liability for flooding, soil moisture distribution and irrigation needs, availability of soil conditioners, such as limestone or peat below the soil or in the proximity, porosity and permeability, possibility of hydraulic contact with underlying aquifer, suitability for the application of liquid manures, likely productivity of the soil, fertilizer requirements, and agroclimatic conditions. Figure 1.5 is a map of the agricultural land-use of part of Togo, West Africa (source: unpublished work of Gu-Konu).

On the other hand, civil constructions (e.g. multistory housing, offices, roads and transport networks, recreational complexes, and water and sewerage works) dominate land-use in urban areas. Consequently, prime consideration is given to the physical and geotechnical properties of the soils and rocks. The urban land-use planner has to take into account two kinds of considerations:

(i) Geohazards, which include the existence of active faults, suscepti- bility to earthquakes, volcanism, slope failure, landslides and sub- sidence, soil liquefaction, river floods and storm surges, coastal ero- sion, etc.

(ii) Considerations related to water supply, sanitation, sewage and dis- posal of liquid effluents and solid wastes: conjunctive use of surface water and groundwater, artificial recharge of aquifers, recycling of waste water, etc.

Land-use planning maps are generally custom-made depending on the needs of the community. The following scenario is generally followed:

Phase 1: preparation of geological, soil, and geohydrological maps, say on a scale of 1:25,000.

Phase 2: Preparation of derivative maps showing slope inclinations, places where the soils need to be drained, where danger of soil erosion exists, areas where drinking water should be protected, existence of historic and religious monuments, etc.

Phase 3: Preparation of derivative maps indicating the possible techno- economic uses for different areas, in terms of projected needs, and as- signing priorities to them.

Phase 4: Conflict map: this is best developed by the preparation of a number of overlays, indicating the conflicting land-use demands for a given area, for the consideration of the planning authority.

Phase 5: Map showing the final decision of the Planning Authority with regard to land-use, after consideration of the techno-socioeconomic aspects of conflicting land-use claims. This would be the final land-use map and the one implemented. Such maps tend to be detailed, say on a scale of 1: 5,000.

Figure 1.6 is the legend of a possible geoscientific land-use map. It indicates the kind of data that is generally needed for the preparation of land-use maps (Archer et al., 1987, p. 198). It should be emphasized that the legend is only indicative; several other land-use parameters may have to be taken into account, depending on the biophysical and cultural envi- ronments of a particular area.

Advances in computer graphics and cartography have now made it possible to prepare the above types of maps quickly and also to visualize how the various scenarios would look on the ground, even though no construction has actually taken place ("virtual reality").

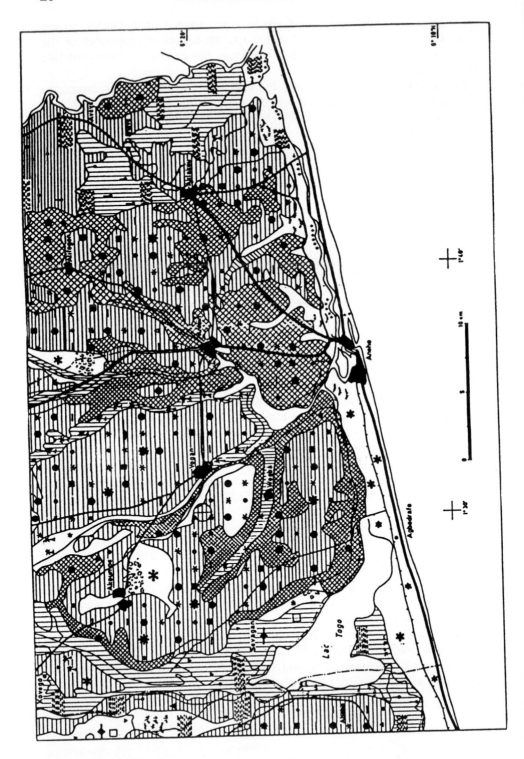

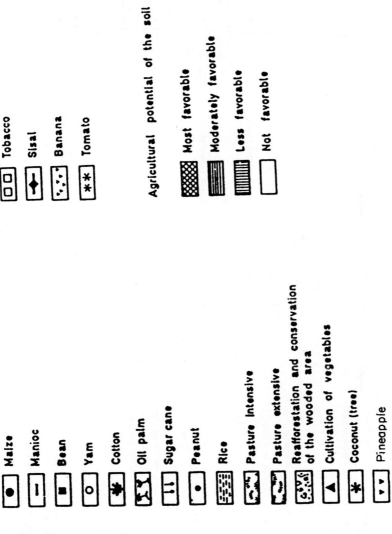

Maize
Manioc
Bean
Yam
Cotton
Oil palm
Sugar cane
Peanut
Rice
Pasture intensive
Pasture extensive
Reafforestation and conservation of the wooded area
Cultivation of vegetables
Coconut (tree)
Pineapple

Tobacco
Sisal
Banana
Tomato

Agricultural potential of the soil

Most favorable
Moderately favorable
Less favorable
Not favorable

Fig. 1.5 Map of agricultural land-use in an area in Togo, West Africa (quoted by Archer et al., 1987, p. 196-197, © UNEP-UNESCO).

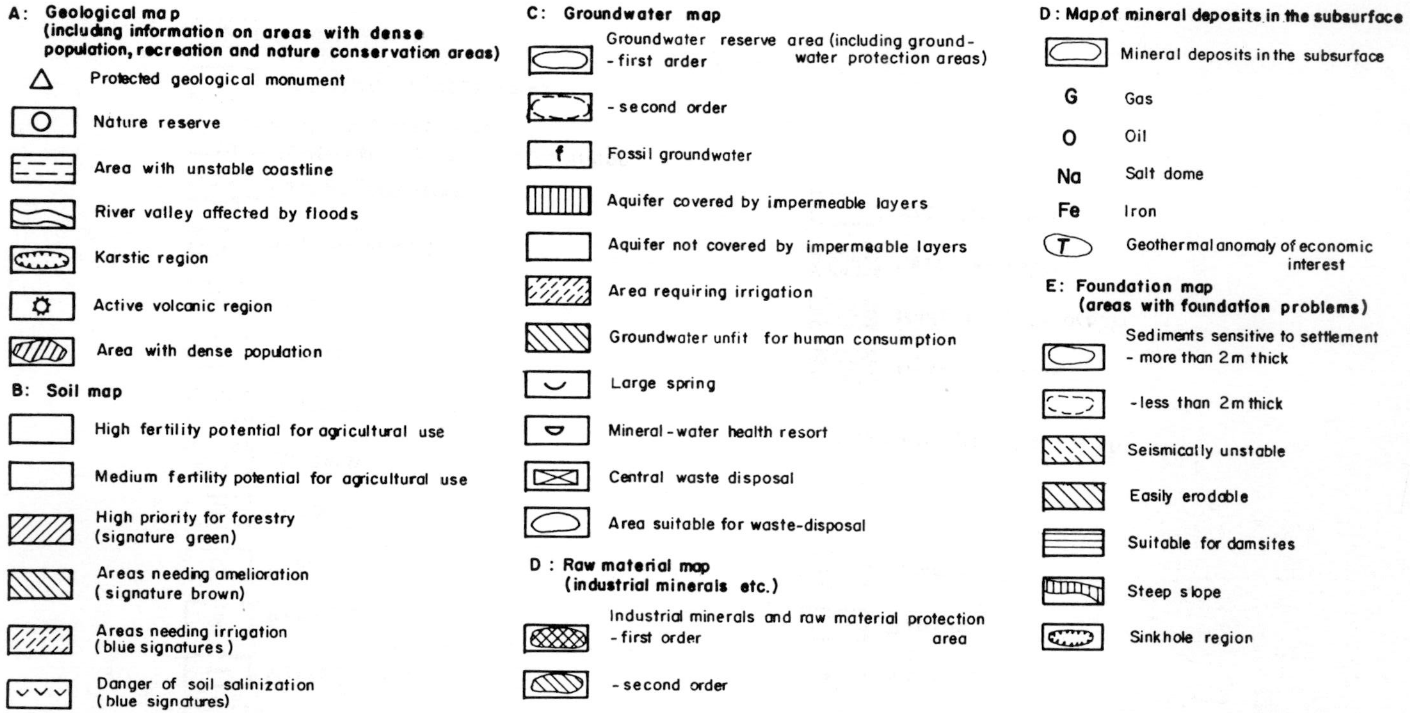

A: Geological map
(including information on areas with dense population, recreation and nature conservation areas)

Protected geological monument

Nature reserve

Area with unstable coastline

River valley affected by floods

Karstic region

Active volcanic region

Area with dense population

B: Soil map

High fertility potential for agricultural use

Medium fertility potential for agricultural use

High priority for forestry (signature green)

Areas needing amelioration (signature brown)

Areas needing irrigation (blue signatures)

Danger of soil salinization (blue signatures)

C: Groundwater map

Groundwater reserve area (including ground-water protection areas)
- first order

- second order

Fossil groundwater

Aquifer covered by impermeable layers

Aquifer not covered by impermeable layers

Area requiring irrigation

Groundwater unfit for human consumption

Large spring

Mineral-water health resort

Central waste disposal

Area suitable for waste-disposal

D: Raw material map (industrial minerals etc.)

Industrial minerals and raw material protection area
- first order

- second order

D: Map of mineral deposits in the subsurface

Mineral deposits in the subsurface

G   Gas

O   Oil

Na   Salt dome

Fe   Iron

T   Geothermal anomaly of economic interest

E: Foundation map (areas with foundation problems)

Sediments sensitive to settlement
- more than 2m thick

- less than 2m thick

Seismically unstable

Easily erodable

Suitable for damsites

Steep slope

Sinkhole region

Fig. 1.6   Legend of a possible geoscientific land-use map (*source:* Archer et al., 1987, p. 198, © UNEP-UNESCO)

The new technique of neural computing is revolutionizing the overlay methodology. Conventional programming is not necessary for neural computing. The neural networks can be taught to perform complex tasks by learning from examples. The neural computers themselves identify the relationships between data sets and recognize patterns and associations; in effect, they build their own "rules". They thus have the ability to learn from experience and to cope with "fuzzy" data. These developments enable the decision-making authorities to appreciate the issues involved, even though they may not possess the technical background for reading the maps directly.

1.8 ANTHROPOGENIC IMPACT ON SOIL ENVIRONMENT

1.8.1 Geoenvironmental Ensembles

Soil environment cannot be considered in isolation for the simple reason that it is an integral part of the geoenvironment. The latter comprises rocks, soils, fluids, gases and organisms. Geoenvironment is linked to and is influenced by, climate, terrain, and vegetal cover.

Human activities influence the geological, physical, chemical, and biochemical processes taking place in rocks and soils, and associated media, in the following manner:

(i) *Rocks*: Change in the composition, structure and geotechnical properties of rocks and overburden;

(i) *Soils*: Creation of new primary soils and modification of soil-forming processes, which affect the bedrock and the groundwater regime;

(iii) *Surface water and groundwater*: These are by far the most dynamic components of the geoenvironment and are most affected by the activities of man. Protection of surface water and groundwater from depletion and pollution (chemical, organic, thermal, mechanical, etc. contamination) constitute the most important environmental task of a community. Microorganisms affect the hydrochemical processes and water quality. For instance, thionine bacteria oxidize hydrogen sulfide to sulfuric acid and render the water more corrosive;

(iv) *Atmosphere*: Atmospheric pollution can penetrate the hydrosphere, biosphere, and lithosphere, and can bring about undesirable changes in climate, soils, vegetation, and water quality.

1.8.2 Technological Considerations

Paul and Ann Ehrlich of Stanford University (USA) gave the following formula regarding the environmental impact of man:

I = P.A.T, where
I = Impact on environment,
P = Population
A = Per capita consumption of resources
T = Technology (environmentally harmful technology used in the consumption of resources, A).

The World Resources Institute (1991) looks at the problem from a different angle. Pollution (understood as environmental degradation) is said to be a function of population, income levels (GDP per capita) and the pollution intensity of production (pollution/GDP).

The above formulations indicate the steps that the industrialized and developing countries need to take in order to minimize the anthropogenic impact of man:

(i) *Control population growth*: The more the population, the greater the environmental impact. Most of the industrialized countries have already achieved stable populations. It is in the developing countries that most of the population growth in the world is occurring. They should therefore strive hard to check the population growth and stabilize it.

(ii) *Control the consumption of resources*: The greater the consumption of resources, the greater the environmental impact. The per capita consumption of resources in the industrialized countries is several times higher than in the developing countries. Industrialized countries should therefore limit their conspicuous consumption and adopt conservation practices. All countries (both industrialized and developing) should recycle wastes to the maximum extent possible.

Recycling of wastes is highly beneficial and cost-effective for the following reasons: (i) cleaning up the surroundings and improvement of hygiene; (ii) reducing potential environmental degradation through conservation (for instance, when paper is recycled, it saves trees from being cut down); (iii) generation of employment; (iv) nobody would object to the garbage being collected and hence the "input" material for waste cycling is virtually free; (v) there is considerable saving of energy (for instance, recycling of aluminum cans needs only 5% of the energy required to make them); and (vi) only simple skills are needed in most cases.

(iii) *Choose technology wisely*: Consistent with economic considerations, both industrialized and developing countries should choose no-waste and low-waste process technologies, in order to achieve minimal environmental impact.

1.8.3 Dimensions of Anthropogenic Impact

It should be self-evident that man has a profound impact on the environment. The impact is caused by mining, various industries, land reclamation and irrigation projects for agriculture, urban development, hydroelectric schemes, waste disposal, including underground storage, construction of roads, tunnels and other civil engineering works, high-pressure injection of liquids into depths of the earth,etc.

The impact of man on geoenvironment is summarized in Table 1.5 (after Archer et al., 1987, p.49).

Table 1.5: Human impact on geoenvironment

Activity	Adverse effects
Mining	Changes in landscape, landslides, subsidence, pollution of water and soil, lowering of groundwater, damage caused by explosions
Civil engineering	Landslides, mudflows, rockfalls, flooding, subsidence, changes in the groundwater level
Industries	Pollution of water, soil, air and biota
Agriculture and forestry	Erosion, landslides, soil salinization, water pollution, flooding
Tourism	Water and soil pollution
Waste disposal	Pollution of water and soil, attack on underlying rocks, thermal effects

Broadly speaking, the environmental degradation in the industrialized countries is largely chemical and these countries are said to be facing a "Chemical Time Bomb". On the other hand, the environmental degradation in the developing countries is largely physical, in the form of soil erosion and desertification. This is obviously an oversimplification, as chemical or biological or physical degradation cannot happen alone, without having an impact on the others. Since 1975, the Geological Survey of Czechoslovakia continuously monitored the amount and chemical composition of atmospheric deposition, runoff, applied fertilizers, harvesting of crops and lumbering of trees, etc. in small representative catchments. This enabled them to study the biogeochemical cycles of elements in relation to rates of weathering and erosion. It was found that acidification increased the rate of chemical erosion, while agriculture increased mostly the rate of mechanical erosion. Chemical erosion increased by a factor of 1.5 due to agriculture and by a factor of 2.9 due to acidification. Agricultural practices increased mechanical erosion by 2.7 times, while acidification due to industrial emissions increased the mechanical erosion by 1.9 times (Paces, 1991).

Just because the developing countries do not have much industrialization does not necessarily mean that they need not worry much about industrial pollution. In reality, a given industry of a given size in a developing country is generally twice as polluting as its analogue in the industrialized countries because the developing countries cannot afford cleaner technologies, and because the environmental standards in those countries either do not exist or cannot be enforced. Whatever industries exist in the developing countries tend to be located near urban centers and discharge their effluents untreated into the municipal sewerage systems. In this kind of situation, even a small industrial unit can cause disproportionately large damage to the environment. For instance, the effluents from a battery reconditioning unit dumped into a stream may contaminate the potable water resources of a large town located downstream. Most likely, neither the battery unit nor the town will have instrumental facilities for monitoring the heavy metal contamination and nobody may even be aware of the damage!

1.9   ECOLOGICALLY SUSTAINABLE USE OF SOIL

Depleted soils mean reduced nutrition to poor farmers. When the soils are degraded, they would have greater susceptibility to drought. The World Bank estimates that field productivity losses due to soil degradation in tropical soils are of the order of 0.5 to 1.5 % of gross national product. Soil erosion could lead to offsite siltation of reservoirs, river-transport channels, and other hydrologic investments.

The evocative concept of "Sustainable Development", first mooted in the famous report (*Our Common Future*, 1987) of the World Commission on Environment and Development, chaired by Madame Brundtland, Prime Minister of Norway, has captured the imagination of the people of the world. Sustainable development is that kind of development that "meets the needs of the present without compromising the ability of future generations to meet their own needs". There has been much debate about the concept of sustainability. It is centered around the appropriate "mix" of the man-made and natural capital that needs to be preserved for future generations, and whether the two kinds of capital are mutually substitutable.

In 1991, the United Nations Industrial Development Organization (UNIDO) reexamined the whole issue and came up with the concept of Ecologically Sustainable Industrial Development (ESID). ESID is defined as those patterns of industrialization that enhance economic and social benefits for present and future generations *without impairing basic ecological processes*. This implies that any industrialization and other human activities that lead to significant degradation of the ecological processes, are deemed to be *ipso facto* unsustainable and unacceptable.

Ecologically sustainable use of soils should satisfy three criteria:

(i) The biosphere should be protected from anthropogenic activities: The protection "includes stabilizing the biosphere in the face of threats from

greenhouse gases and ozone-depleting substances, maintaining the carrying capacity of the natural resource systems (forest, fisheries, agricultural land), and protecting the absorptive (assimilative) capacity of air, water and soil from emissions and waste discharges".

(ii) It must make the most efficient use of man-made and natural capital: When fertilizers are used to improve soil productivity, care must be taken to ensure that the fertilizers are not leached into stream waters which are used for drinking purposes downstream. Similarly, when pesticides are used to protect the crops from pests, they should be so chosen that either they are ecologically friendly or biodegradable or have a short residence time in the soil or do not leave much residue on the agricultural product concerned.

(iii) It must be socially equitable: This applies as much to industrialized versus developing countries, as to rich versus poor people in a given country. Environmental degradation makes poor people poorer — degraded land means less food, water, and fuel. The poor suffer the consequences of degradation more. If your neighbor's farm is getting gullied because of unsustainable agricultural practices, your farm cannot escape degradation (if your neighbor's house is on fire, it is not just his problem alone !). Just as the industrialized countries are asked to assist the developing countries to stop the global environmental degradation in their own larger self-interest, the well-to-do members of a community should assist the poorer members in ameliorating the environmental situation.

1.10 LEISA

Innovative ideas have been developed to increase the agricultural productivity of soils, through Low External Input, Sustainable Agriculture (LEISA).

Sustainable agriculture has been defined as the "successful management of resources of agriculture to satisfy changing human needs while maintaining and enhancing the quality of life and the environment and conserving natural resources". Such a management has to be ecologically sound, economically viable, socially just, humane and adaptable. Time-tested, socially acceptable and scientifically sound indigenous soil management practices are incorporated into the framework of activities.

The Participatory Technology Development (PTD) constitutes the core of the new approach. It has the following components:

(i) Getting started — networking and making inventories,

(ii) Looking for things to try — developing the research agenda,

(iii) Designing experiments — building on local experimental capacity,

(iv) Trying things out — implementing and evaluating experiments,
 (v) Sharing the results — communication, dissemination, and training,
(vi) Keeping up the process — embedding the local technology.

Soil management is inseparable from water management. In some situations, it may be convenient to collect precipitation and store it in ponds for use in irrigation. In drought-prone areas, it is necessary to conserve moisture by various physical and agronomic methods. Strategies have been developed for concurrent production and conservation, whereby control of soil erosion and maintenance of organic matter, moisture and physical properties of the soil, go hand in hand with increase in the production of food, fuel, and fodder.

Dynamics of Soil Processes and Soil Environments

All things come and go. A star melts as surely as a snow-flake...only
to come again in some other time and place-

—Walt Whitman

2.1 MINERALOGY OF SOILS

Soils contain two broad groups of minerals: Primary silicate minerals, such as quartz, feldspars, mica, amphibole, pyroxene and, very rarely olivine, which are directly inherited from the parent rocks, and a large number of the so-called secondary minerals which come into existence in the course of soil formation.

Table 2.1 lists the common soil minerals (compilation after Sposito, 1989, p.8). The phyllosilicates related to clay minerals are described in Table 2.2. The structure and chemical formula of important clay mineral groups are given in Table 2.3 (*source*: Sposito, 1989, p. 33).

In the process of soil formation, some trace elements get coprecipitated with secondary soil minerals and organic matter (Table 2.4; *source*: Sposito, 1989, p. 12). This information enables us to predict on the basis of the abundance of soil minerals, which trace elements are likely to be present in a given soil.

Since soils are open systems, there is continual input and output of water (precipitation, drainage, evapotranspiration), biomass and energy (radiation) in them. Consequently, the soils constantly undergo change with the passage of time. This change is reflected in the mineralogy of the clay fraction. The three stages of weathering (as determined by Jackson and Sherman) have characteristic chemical and physical conditions in the soil and characteristic development of minerals in the soil clay fraction, as follows:

Early stage: Very low content of water and organic matter; very limited leaching — gypsum, carbonates, Fe(II)-bearing micas, feldspars.

Table 2.1 Common Soil Minerals

Mineral	Presence and significance
Quartz	Abundant in sand and silt
Alkalic and calcic feldspars	Abundant in soils that are not leached intensively
Micas	Source of K in most temperate-zone soils
Amphibole	Easily weathered to clay minerals and oxides
Pyroxene	Easily weathered
Olivine	Easily weathered. Almost never found in soils.
Epidote, tourmaline, zircon, rutile	Highly resistant to chemical weathering; used as "index" minerals in pedologic studies
Kaolinite, smectite, vermiculite, and chlorite	Clay products of weathering; source of exchangeable cations in the soils
Allophane, imogolite	Abundant in soils derived from volcanic ash deposits
Gibbsite	Abundant in leached soils
Goethite	Most common Fe oxide in soils
Hematite	Abundant in warm regions
Ferrihydrite	Common in hydromorphic soils
Birnessite	Most abundant Mn oxide
Calcite	Most abundant carbonate
Gypsum	Abundant in arid regions

Intermediate stage: Limited leaching, retention of Na, K, Ca, Mg, Fe(II) and silica; silicates easily hydrolyzed, flocculation of silica and transport of silica into the weathering zone — quartz, dioctahedral mica/illite, vermiculite/chlorite, smectites.

Advanced stage: Intensive leaching, fresh water; removal of Na, K, Ca, Mg, Fe(II), and silica; oxidation of Fe(II); acidic compounds; low pH; dispersion of silica; Al-hydroxy polymers — kaolinite, gibbsite, iron oxides (goethite, hematite), titanium oxides (anatase, rutile, ilmenite).

The rate and direction of weathering processes are critically dependent upon the concentration of silicic acid in the soil solution. As the concentration of silicic acid decreases due to leaching, the soil clay fraction changes from illite to kaolinite to gibbsite.

2.2 CHEMICAL COMPOSITION OF SOILS

The ten most abundant elements in the soils (in the order of abundance) are: O > Si > Al > Fe > C > Ca > K > Na > Mg > Ti, whereas in crustal rocks, the order of abundance is: O > Si > Al > Fe > Ca > Mg = Na > K > Ti > P.

Table 2.2   Classification of phyllosilicates related to clay minerals
(*source:* Glossary of Soil Science Terms, 1987, p.23)

Type	Group (x = charge per formula unit)	Subgroup	Species*
1 : 1	Kaolinite-serpentinite x = 0	Kaolinite	Kaolinite, halloysite (7 Å) halloysite (10 Å)
		Serpentines	Chrysotile, lizardite antigorite
2 : 1	Pyrophyllite-talc, x = 0	Pyrophyllite Talcs	Pyrophyllite Talc
	Smectite x = 0.25 – 0.6	Dioctahedral smectites	Montmorillonite, bedellite, nontronite
		Trioctahedral smectites	Saponite, hectorite, sauconite
	Vermiculite x = 0.6 – 0-9	Dioctahedral vermiculite	Dioctahedral vermiculite
		Trioctahedral vermiculite	Trioctahedral vermiculite
	Mica x = 1	Dioctahedral micas	Muscovite, paragonite
		Trioctahedral micas	Biotite, Phlogopite
	Brittle mica x = 2	Dioctahedral brittle micas	Margarite
		Trioctahedral brittle micas	Clintonite
2 : 1 : 1	Chlorite x variable	Dioctahedral chlorites (4 - 5 octahedral cations per formula unit)	
		Trioctahedral chlorites (5-6 octahedral cations per formula unit)	Pennine, chlinchlore, prochlorite

*Only a few examples are given

The soil enrichment factor of nutrient elements is obtained by dividing the mean content of the element in the soil by the mean content of the same element in the crustal rocks (Table 2.5; *source*: Sposito, 1989, p.6).

Table 2.3   Chemical composition of important clay mineral groups

Group	Layer type	Layer charge (x)#	Typical chemical formula
Kaolinite	1 : 1	< 0.01	$[Si_4]Al_4O_{10}(OH)_8 \cdot nH_2O$ (n = 0 or 4)*
Illite	2 : 1	1.4 - 2.0	$Mx[Si_{6.8}Al_{1.2}]$ $Al_3 Fe_{0.25} Mg_{0.75}$ $O_{20}(OH)_4$
Vermiculite	2 : 1	1.2 - 1.8	$Mx[Si_7 Al]$ $Al_3 Fe_{0.5} Mg_{0.5}$ $O_{20}(OH)_4$
Smectite†	2 : 1	0.5 - 1.2	$M_x[Si_8]$ $Al_{3.2} Fe_{0.2} Mg_{0.6}$ $O_{20}(OH)_4$
Chlorite	2 : 1, with hydroxide interlayer	variable	$(Al(OH)_{2.55})_4$ $[Si_{6.8}Al_{1.2}]$ $Al_{3.4} Mg_{0.6}$ $O_{20}(OH)_4$

$x = 12 - a - b - c$, where a, b and c are the stoichiometric coefficients of Si, octahedral Al, and Fe (II) respectively.
* n = 0 for kaolinite and n=4 for halloysite.
M = Monovalent interlayer cation.
† principally montmorillonite

Table 2.4   Soil minerals and coprecipitated trace elements

Soil mineral	Coprecipitated trace elements
Fe and Al oxides	B,P,V,Mn,Ni,Cu,Zn,Mo,As,Se
Mn oxides	P,Fe, Co,Ni,Zn,Mo,As,Se,Pb
Ca carbonates	P,V,Mn,Fe,Co,Cd
Illites	B,V,Ni,Co,Cr,Cu,Zn,Mo,As,Se,Pb
Smectites	B,Ti,V,Cr,Mn,Fe,Co,Ni,Cu,Zn,Pb
Vermiculites	Ti,Mn,Fe
Organic matter	Al,V,Cr,Mn,Fe,Ni,Cu,Zn,Cd,Pb

The very high enrichment of nitrogen and carbon and moderate enrichment of sulfur and boron in the soils in relation to the crust, are an obvious consequence of the operation of the biologic processes in the soils, as manifested in the form of soil organic matter and soil organisms.

The solid matter constitutes one-half to two-thirds of the soil volume. Normally, the bulk of the soil mass is composed of inorganic compounds, though in peaty soils the organic matter may account for up to 90% of the mass. The inorganic solid phases, particularly the secondary minerals such as clays, tend to be in a metastable state. The soil organic compounds have complex structures.

Table 2.5 Mean content (in mg kg^{-1}) of essential elements in the soil and crustal rocks, and the soil enrichment factor (EFs)

Element & At. no.	Soil	Crustal rocks	EF$_s$
B (5)	33	10	3.3
C (6)	25,000	480	52
N (7)	2,000	25	80
O (8)	490,000	474,000	1.0
Mg (12)	9,000	23,000	0.39
P (15)	430	1,000	0.43
S (16)	1,600	260	6.2
Cl (17)	100	130	0.77
K (19)	15,000	21,000	0.71
Ca (20)	24,000	41,000	0.59
Mn (25)	550	950	0.58
Fe (26)	26,000	41,000	0.63
Cu (29)	25	50	0.50
Zn (30)	60	75	0.80
Mo (42)	0.97	1.5	0.65

Humic substances are the dark, microbially transformed materials present in the soil organic matter. They can be broadly divided into humic acids ($C_{187} H_{186} O_{89} N_9 S$) and fulvic acids ($C_{135} H_{182} O_{95} N_5 S_2$). It is instructive to compare the chemical composition of the humic and fulvic acids with the average C/N/P/S *molar* ratio of soil organic matter (278:17:1:1) and the average chemical composition of aquatic plants ("Redfield formula"— $C_{106} H_{263} O_{110} N_{16} P$). The humic and fulvic acids are depleted in N and enriched in S, relative to organic matter. They are enriched in C and depleted in H and N, in relation to living organisms and biomolecules (as represented by the "Redfield formula"). The depletion of N and their aromatic character render the humic and fulvic acids more capable resisting microbial attack.

2.2.1 Trace Elements in the Soil

As should be expected, the distribution of trace elements in a soil profile depends upon the trace element composition of the parent material and the nature and duration of the soil-forming processes. The trace elements commonly found in the soils have an enormous range — for instance, the average concentration of Ti (about 10,000 ppm) is about 10^5 times higher than gold (0.1 ppm) (Fig. 2.1; *source:* Andrews-Jones, 1968).

The relative mobilities of different trace elements (essential or toxic) in the surficial environments are summarized in Table 2.6 (*source:* Andrews-Jones, 1968).

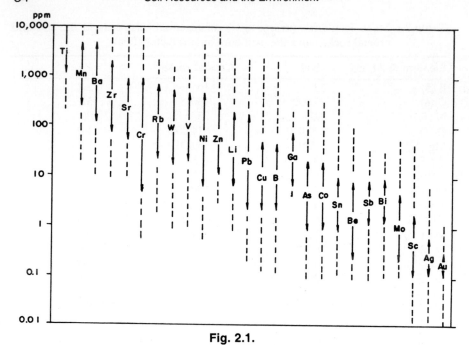

Fig. 2.1.

Table 2.6 Mobilities of essential and potentially toxic elements under various surficial environments (*source:* Andrews-Jones, 1968)

Relative mobilities	Oxidizing	Acid	Neutral-alk.	Reducing
Very high	I	I	I Mo U Se	I
High	Mo U Se F Ra Zn	Mo U Se F Ra Zn Cu Co Ni Hg	F Ra	F Ra
Medium	Cu Co Ni Hg As Cd	As Cd	As Cd	As Cd
Low	Pb Be Bi Sb Tl	Pb Be Bi Sb Tl Fe Mn	Pb Be Bi Sb Tl Fe Mn	Fe Mn
Very low or immobile	Fe Mn Al Cr	Al Cr	Al Cr Zn Co Cu Ni Hg	Al Cr Mo U Se Zn Co Cu Ni Hg As Cd Pb Be Bi Sb Tl

Note: Roman indicates essential trace elements; bold indicates potentially toxic

Trace elements in the soil play an important part in plant nutrition. A trace element is any chemical element whose mass concentration in the solid phase is less than 100 mg kg^{-1} (100 parts per million or ppm). A trace element may be a nutrient element or a toxic element. Either way, it is necessary to understand as to with which primary mineral a given trace element is associated, the pathways of the trace element in the process of weathering and the secondary mineral with which the trace element may be coprecipitated.

This can be illustrated with the examples of a nutrient element (Zn) and a toxic element (Cd).

Zinc is a trace element and a micronutrient in the soil. Zinc occurs in the primary minerals in the form of sulfide inclusions in silicates, and through isomorphic substitution for Fe and Mg in olivines, pyroxenes and amphiboles, and for Fe or Mn in oxides. In the soils, it gets coprecipitated with Fe and Mn oxides and organic matter. The ability of a soil to provide zinc to plants depends upon the rate at which zinc gets solubilized (through the operation of factors such as pH, redox potential, and water content) and becomes available to the plant. This consideration underlines the importance of the preservation of soil humus to ensure adequate nutrition of zinc.

Cadmium is a toxic trace element. It is associated with primary minerals in the form of sulfide inclusions and in isomorphic substitution for Cu, Zn, Hg and Pb in sulfides. In the secondary environment of soils, it gets coprecipitated with calcium carbonates. This suggests that liming is an effective way to sequester cadmium in the soil and prevent it from reaching the plants.

2.3 SOIL ORGANIC PROCESSES

2 3.1 Soil Biota

Soil is teeming with life — figuratively and literally. One kilogram of soil may contain 500 billion bacteria, 10 billion actinomycetes, and about a million fungi. The combined length of roots of a single plant may reach an incredible length of 600 km in the top meter of the soil profile (Sposito, 1989, p.42). Almost any bacterium known to man can be found in the soil. Some may exist for short periods. Some organisms, restricted in range live, grow, and attain high population levels (about 10^4 to 10^5 per gram of soil).

The following account of soil biology has largely been drawn from Dindal (1990).

Soil microorganisms, such as bacteria, actinomycetes, fungi, algae and protozoa, are involved in "decomposition of organic matter and production of the humus, cycling of nutrients, energy and elemental fixation, soil metabolism, and the production of complex compounds that cause soil aggregation" (Dindal, 1990). These organisms are characterized by high surface area-to-volume ratio. They are, in effect, living tissues — pools of chemical elements, compounds and energy. Some soil microbes have a symbiotic relationship with higher plants as nitrogen fixers.

Soil invertebrates work in tandem with soil microorganisms, and serve as regulators of microbial processes through comminution, translocation, defecation, and inoculation of microbial propagules. They thus play an important role in determining the biological, chemical, and physical characteristics of the soil ecosystem. Through their burrowing and complex feeding activities, the soil invertebrates bring about mechanical blending of the soil components, and modification of the rates of humification. The deposition of their fecal pellets affect the size and frequency of soil aggregates. Soil invertebrates cause the enhancement of soil aeration and nutrient and water-holding capacities. This,however, has an adverse consequence — the bulk density of the soil gets reduced and the soil becomes more susceptible to erosion by water and wind.

Actinomycetes are filamentous, with branched growth habit. The most abundant series are: *Streptomyces* spp., *Nocardia* spp., and *Micromonospora* spp. They are well known for their ability to produce antibiotics, such as streptomycin and tetracycline. Actinomycetes are of considerable ecological importance. Their abundance is dependent upon moisture content, aeration, organic material, pH, rhizosphere and vegetation, soil depth and soil type. Actinomycetes are resistant to desiccation; consequently, lower soil moisture content favors their survival and growth. Soils that are rich in organic matter or humus have greater numbers of actinomycetes, which occur predominantly in subsurface soil layers. Many actinomycetes are pH-sensitive; they constitute a dominant component of microbial populations under alkaline conditions but are intolerant of acid pH below 5.0.

The distribution of fungi in soils depends on the nature and distribution of organic substrates, soil moisture and temperature, and soil depth. Fungi exist in soils, as spores, dormant propagules, or hyphae in various physiological states. The hyphae are particularly effective in the decomposition of organic matter. They have large surface area/ volume ratios, and are capable of ramifying over and penetrating into organic substrata for the absorption of soluble nutrients.

Protozoa connote a level of animal organization rather than a "natural" group of organisms. Free-living protozoa are ubiquitous in the decomposing plant material in the surface layers of the soil. Soil moisture is the most significant abiotic factor affecting the distribution of protozoa in the soil. Evaporation or evapotranspiration increases the relative concentration of salts and subjects the protozoa to osmotic stress. On the other hand, heavy precipitation may cause dilution stress and may create waterlogging and anaerobic conditions. Another kind of stress to which protozoa may be subjected is high concentrations of carbon dioxide that sometimes build up due to the decomposition of detrital matter and root respiration.

2.3.2 Soil Biochemical Processes

Soil microorganisms profoundly influence the chemical processes in the soil. They not only catalyze the oxidation-reduction reactions in the soil, but also exude organic acids which play a critical role in soil acidity and the cycling of trace elements. Table 2.7 (*source:* Sposito, 1989, p. 43) lists the five aliphatic organic acids produced by microbial activity in the *rhizosphere*, which is the local soil environment most affected by plant roots. These acids have a common feature: they contain the *carboxyl* organic unit, COOH, which readily dissociates into proton (H^+) and carboxylate anion (COO^-) in the normal range of soil pH (see the third column in Table 2.7). The proton thus released attacks and decomposes the soil minerals and the carboxylate ion may form soluble complexes by combining with metal cations released by mineral weathering. In the soil solution, the total concentration of organic acids ($0.01–5$ mol m^{-3}) is significantly higher than the concentration of trace metals (<1 mmol m^{-3}). These organic acids in the soil have very short lifetimes, of the order of a few hours, but they are produced continually by microbial activity.

Table 2.7 Common aliphatic organic acids in soil

Aliphatic organic acids	Chemical formula	pH$_{dis}$*
Formic acid	HCOOH	3.8
Acetic acid	CH$_3$COOH	4.8
Oxalic acid	HOOCCOOH	1.3
Tartaric acid	H O | H HOOCCCCOOH H | O H	3.0
Citric acid	COOH H | H HOOCCCCCOOH H | H O H	3.1

* pH$_{dis}$ refers to the pH value at which most acidic carboxyl groups have a 50% probability of being dissociated in aqueous solution.

Besides these aliphatic acids, the soil solutions contain some *aromatic* acids also. As is well known, the benzene ring is the fundamental structural unit of aromatic acids. A variety of aromatic acids develop when carboxyl (COOH) or hydroxyl (OH) groups are bonded to the benzene ring. The concentration of aromatic acids in the soil solution is of the order of 0.05 – 0.3 mol m^{-3}.

Soil solutions also contain a large variety of other organic compounds, such as amino acids, peptides, various polymers, monosaccharides and polysaccharides, phenols, lignin, organophosphorus and organosulfur compounds, etc. (for details, the reader is referred to the excellent volume, *Humus Chemistry* by Stevenson, 1982).

2.4   SOIL HUMUS AND CATION EXCHANGE CAPACITY

The dark-colored, well-decomposed organic matter in the soil is called the *humus*. Humus plays an important role in the formation and stabilization of soil aggregates, control of soil acidity, cycling of nutrient elements and detoxification of toxic substances such as heavy metals and pesticides.

Humus may exist as a colloid or as a coating on mineral surfaces. When humus comes into contact with organic substances such as pesticides, fertilizers, green manure, etc., it can immobilize them, and sometimes detoxify them. Thus, when inorganic fertilizers are applied in combination with humus-bearing wastes (such as brown coal sludges, brown coal dust), the humus will hold onto the nutrient elements, such as phosphorus and sulfur, and prevent their ready leaching.

The role of humus has another dimension. Soluble humus compounds, such as fulvic acid, can form complexes with organic compounds, which may then move along with percolating water. Thus, a pesticide deposited on the soil surface does not stay there; it can move and contaminate the groundwater resources.

Humus existing as coatings on soil minerals plays an important role in the cycling of chemical elements and in the formation of aggregates. Where humus is bound to clay minerals, it can resist degradation, while at the same time providing a reactive surface to dissolved solutes in the soil solution. Where the clays are not coated with humus or where the humus coating has been partly dissolved, soluble anions produced by microbial action may interact with cations adsorbed on clays. Thus soil minerals serve both as a substrate for humus and a source of metal ions with which to form soluble humus complexes (Sposito, 1989, p. 59).

In totality, humus is chemically very complex. It is generally studied through the separation and characterization of its constituent substances. Soil is treated with a 500 mol m^{-3} NaOH solution. The extract containing the soluble organic material is brought to pH 1 with concentrated HCl. *Humic* acid is the precipitate formed after this acidification and the *fulvic* acid is the material remaining in solution.

The average chemical composition of humic acid (in g kg^{-1}) is: C (560), H (47), N (32), S (8), O (355) COOH (3.6), phenolic OH (3.1). Its molecular formula is: $C_{187} H_{186} O_{89} N_9 S$. The average chemical composition of fulvic acid (in g kg^{-1}) is: C (457), H (54), N (21), S (19), O (448), COOH (8.2), and phenolic OH (3.0). Its molecular formula is: $C_{135} H_{182} O_{95} N_5 S_2$. It may be noted that the fulvic acid contains more carboxyl groups and hence is capable of providing a greater number of protons than the humic acid. This has important implications for cation exchange in soils.

2.4.1 Cation Exchange Capacity

A continuous exchange occurs between the ions in solution and ions held on the exchange sites of minerals and organic matter. The ion exchange capacity of a soil is its ability to hold and exchange ions. This involves mostly cations, though in a few cases anions of arsenates and selenates may be involved. The *Cation Exchange Capacity* (CEC) is given by the total potential of a soil to absorb cations, which is largely determined by the content of organic matter and clays. Various elements form humic complexes with the organic components of the soil. The high ion-exchange capacity of the clays arises from their layer-lattice structures and large surface areas. The greater the charge of the ions, the greater their sorption — thus, $M^{3+} > M^{2+} > M^+$. For cations of the same charge, the ionic size of the hydrated cation determines the relative degree of sorption. The capacity of the clays to concentrate various metals is the highest for montmorillonite (2:1 expanding clay), followed by vermiculite (2:1 limited expanding clay), illite (2:1 non-expanding clay), chlorite (2:2 clay), with kaolinite (1: 1 clay) having the least capacity. The generally high fertility of volcanic soils is a consequence of their montmorillonite content with high sorption capacity.

In the SI system, CEC is expressed in terms of moles of charge per kilogram of adsorbent (mol_c kg^{-1}) (previously milliequivalents/100 g). Typical CEC values (in $mols_c$ kg^{-1}) for soil colloidal constituents are as follows: soil organic matter: 150 — 300; kaolinite (clay): 2 — 5; illite (clay): 15 — 40; montmorillonite (clay): 80 — 140; vermiculite (clay): 150; hydrous oxides of Fe, Al and Mn: 4.

CEC of a clay (e.g. smectite) depends largely on its layer charge,

$$CEC = (x/M_r). 10^5 \qquad (2.1)$$

where x is the layer charge and M $_r$ is the relative molecular mass of the clay mineral.

Sand and silt are relatively larger particles with smaller surface areas. Their contribution to the CEC of a soil is negligible. On the other hand, the specific surface of the clays is very large, because the platy particles are extremely fine (about 10 Å). For instance, the specific surface of 1 kg of clay may be of the order of about 500,000 m^2 (i.e., the area of about 1,000 soccer fields !). Where the content of organic matter is low, the CEC of a soil is almost wholly attributable to clays (vide Exercise 2).

The molecular structures of the entire range of humic substances are not yet fully understood. The important functional groups in humic substances, particularly those which are most reactive with protons and metal cations, are given in Table 2.8 (*source:* Sposito, 1989, p. 51).

Table 2.8   Important functional groups in soil humus

Functional group	Structural formula
Carboxyl	$\begin{array}{c} O \\ \parallel \\ -C-OH \end{array}$
Carbonyl	$\begin{array}{c} O \\ \parallel \\ -C- \end{array}$
Amino	$-NH_2$
Imidazole	Aromatic ring NH
Phenolic OH	Aromatic ring OH
Alcoholic OH	$-OH$
Sulfhydryl	$-SH$

The cation exchange involving the dissociable protons on soil humus and a cation (such as Ca^{2+}) in the soil solution, can be written in the form of the following equation:

$$SH_2 \text{ (s)} + Ca^{2+} \text{ (aq)} = SCa(s) + 2H^+ \text{ (aq)} \qquad (2.2)$$

where SH_2 represents the amount of humus (S) bearing 2 mol of dissociable protons, and SCa is the same amount of humus bearing 1 mole of exchangeable Ca^{2+}. The symbol, S^{-2}, would represent an amount of humus bearing 2 mol of negative charge that can be neutralized by cations drawn from the soil solution (Sposito, 1989, p. 52).

The CEC of soil humus has been defined as the maximum number of moles of positive charge dissociable per unit mass of humus under given conditions of temperature, pressure, soil solution composition, and humus concentration.

CEC in humus is usually determined on the basis of the moles of proton exchanged in the reaction:

$$2 \text{ SH (s)} + Ba^{2+} \text{ (aq)} = S_2Ba(s) + 2H^+ \text{ (aq)} \qquad (2.3)$$

where Ba^{2+} ions are supplied in a 100 mol m^{-3} Ba(OH)$_2$ solution. The CEC of colloidal humic acids is typically in the range of 4 to 9 mol$_c$ kg^{-1}.

2.5   SOIL–AIR INTERACTION

Between one-third to two-thirds of the soil volume is usually composed of the fluid phases of soil air and soil water. The chemical components of the soil air are broadly similar to those of the atmospheric air, though their relative abundances (in mL L^{-1}) are different.

In the case of O_2, the atmospheric air contains 209 mL L^{-1}, whereas well-aerated soils contain 180 — 205 mL L^{-1}. This figure may go down to 100 mL L^{-1} in the subsurface soil, or even to 20 mL L^{-1} in irrigated and waterlogged soils. The atmospheric air contains 0.335 mL L^{-1} (335 ppm) of CO_2. The soil air contains much larger amounts of CO_2 — normally 3-30 mL L^{-1}, but it may go up to 100 in soil at a depth of one metre, in the vicinity of plant roots, and in flooded soils. The atmospheric air contains N_2 (781 mL L^{-1}), and Ar (9.3 mL L^{-1}). The soil air contains NO, N_2O, NH_3, CH_4 and H_2S, etc. produced by microorganisms under anaerobic conditions.

The very high CO_2 content of the soil air vis-à-vis the atmospheric air has profound implications with regard to soil acidity and carbonate chemistry.

Soil water holds various solids and gases in a dissolved form as a soil solution. Soil water exists mainly as a condensed phase and the concentration of water vapor in the soil can approach 30 mL L^{-1} in a wet soil. The soil solution always contains electrolytes, i.e., ion-forming chemical elements such as K (K^+) and Cl (Cl^-).

The cycling of elements in the soil environment takes place through the equilibration of gases between the soil air and the soil solution. The equilibrium is governed by *Henry's law:*

$$K_H = [A \text{ (aq) }]/P_A \qquad (2.4)$$

where K_H is "Henry's law constant", expressed in units of mol m^{-3} atm^{-1}, [A] is the concentration of gas A in the soil solution (mol m^{-3}) and P_A is the partial pressure of A in soil air (atm.).

The values of "Henry's law constant" (in mol m^{-3} atm^{-1}) for 25°C for the common soil gases are as follows: CO_2 (34.06), CH_4 (1.50), NH_3 (5.76 × 10^4), N_2O (25.55), NO (1.88), O_2 (1.26), SO_2 (1.24 × 10^3) and H_2S (1.02 × 10^2) (data from Stumm and Morgan, 1981).

2.6   SOIL-WATER INTERACTION

All soils have voids or interstitial spaces, their extent depending on the composition of the soil. After precipitation, flooding or irrigation, these interstitial spaces get filled with water, temporarily or permanently. When this happens, the soil is said to be *saturated*. The *saturation capacity* of a soil is defined as the maximum amount of water a soil can hold when saturated. It should, however, be borne in mind that it is physically not possible for all the voids to be completely filled, for the simple reason that some air is inevitably trapped.

2.6.1   Soil Moisture

Soil moisture refers to the amount of water present in the soil above the watertable (Fig. 2.2). Water is held in the soil with varying degrees of tension, ranging from water which is free to flow, to water that is held firmly on the surfaces of the soil particles. On the basis of the process of

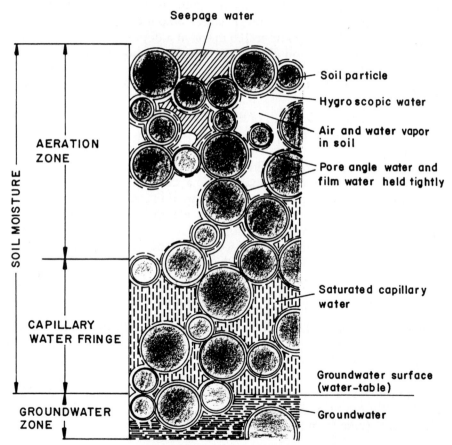

Fig. 2.2   Various dimensions of the soil moisture
(*source:* Chris Barrow, 1987, p. 84, © Longman's)

physical movement involved, soil moisture may be classified into three types: gravity, capillary, and hygroscopic. Gravity (or vadose) water is that water which is present in the soil above the watertable and which drains away under the influence of gravity. Sandy soils take less than a day to drain while this process may take about four days in the case of clay soils. Complete drainage is not possible and considerable amounts of moisture are invariably left behind.

Capillary water moves by capillary action — it "creeps" above the watertable to a height determined by the texture and composition of the soil. The capillary fringe refers to this zone. Where the watertable is shallow, it may happen that the capillary fringe may extend right up to the surface of the soil. The capillary water which thus moves upward evaporates at the surface, leaving behind the dissolved salts that the water may contain or may have leached from the soil in the process of upward movement.

Hygroscopic water refers to the film of water that adheres to the soil particles by molecular attraction. This moisture is virtually inaccessible to plants because of the firmness with which it is held by the soil particles. However, some *xerophytes* (plants that can survive in arid conditions) are known to be able to extract such water.

Field capacity or Field Moisture Capacity of a soil refers to the maximum amount of water remaining in the soil after the gravity water has drained away. The term has been defined in various ways. Field Capacity has been defined by Carruthers and Clark as the quantity of water held at a particular suction pressure forty-eight hours after wetting. The parameter was to be expressed in mm. According to FitzPatrick (1993, p.232), Field Capacity is to be expressed as a percentage of the oven-dry soil. Hillel (1987, p. 35) defines the Field Capacity as the water content of specified volume of soil measurable two days after thorough irrigation, expressed as fractional volume (%). As the draining of soils takes place continuously and unevenly, there is no constancy in field capacity. The term has become largely obsolete.

The relationship between soil type and available moisture is given in Table 2.9 (*source:* Stern, 1979, p.85).

Table 2.9 Soil texture and available water

Soil type	Available water, %	Available water, mm^{-1}
Fine sand	2 - 3	30 - 50
Sandy loam	3 - 6	40 - 150
Silt loam	6 - 8	60 - 120
Clay loam	8 - 14	90 - 210
Clay	13 - 20	190 - 300

Plants growing on the soil will extract moisture as long as it is available. When moisture is no longer available, the plants will wilt and die (wilting point). Fig. 2.3 shows the relationship between the soil moisture characteristics, soil texture, field capacity and wilting point. It may be noted that the wilting point increases with fineness of texture — low for sand and high for clay. The field capacity increases from sand to silt loam but remains unchanged after that.

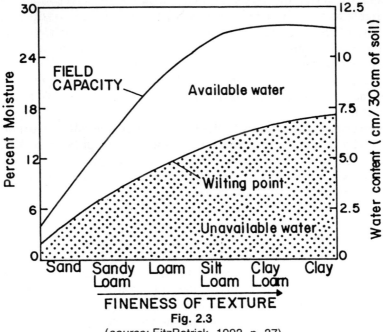

Fig. 2.3
(*source:* FitzPatrick, 1993, p. 37)

Water moves in the soil in three ways: (i) saturated flow: when water is moving through a soil in which all the pores are filled with water. Saturated flow may be in any direction, vertically downward or laterally; (ii) unsaturated flow: when water moves from pore to pore, in a situation wherein only some pores are filled with water. Unsaturated flow takes place both in response to gravity (vertical or lateral movement) as well as to moisture gradient (capillary movement upward); and (iii) vapor transfer: movement of water as a vapor within the soil and from soil to atmosphere. The rate of movement is determined by the relative humidity, temperature gradient, porosity and permeability, and the degree of saturation.

There are two demands upon the soil moisture — evaporative demand of the atmosphere and water needs of the crop plant. Whereas the evaporative demand is continuous, the supply of water by natural precipitation is sporadic. Under the circumstances, the crop has to rely most of the time on the limited reserves of water present in the soil pores. The water economy of some plants is so delicate that if the available soil moisture is inadequate, they become less productive or may even die.

2.6.2 Soil Solutions

The aqueous liquid phase in the soil constitutes the soil solution. Since soil is an open system, the composition of the soil solution is influenced by flows of matter and energy between the soil, atmosphere, biosphere, and hydrosphere.

Complexation is a process whereby an ion acting as a central group attracts other ions and forms associations with them. The associated ions are called *ligands*. The complex may have a positive or negative charge, or it may be neutral. Examples are: Si^{4+}, Al^{3+} and $(CO_3)^{2-}$ acting as a central group to form aqueous complexes, $Si(OH)_4^0$, $Al(OH)_2^+$ and HCO_3^- respectively, with OH^- or H^+ serving as a ligand. The term ligand may also be applied to cations coordinated to an anion, as in the complex, $H_2PO_4^-$. A complex is called a *chelate*, if two or more functional groups of a single ligand are coordinated to a metal cation — for instance, the complex between the metal cation Al^{3+} and citric acid involving two COO^- groups and one COH group, $[Al(COO)_2^- COH(CH)_2 COOH]^+$, is a chelate.

A normal soil solution may contain 100-200 different soluble complexes. Table 2.10 (*source:* Sposito, 1989, p. 69) lists the principal chemical species in the soil solutions in acid soils and alkaline soils. It may be noted that for a given element, acid soils tend to contain free metal cations and protonated anions, whereas alkaline soils contain carbonate or hydroxyl complexes.

A given chemical element may exist in several chemical forms, depending on their stability constants. On the basis of the measured total concentrations of metals and ligands along with the pH values, the conditional stability constants for all possible complexes, and the expressions for mass balances of each constituent, it is possible to compute the speciation of a particular constituent.

2.6.3 Soil Mineral Solubility

When water molecules enter a dry soil, they will invade the soil minerals through microfractures and structural imperfections. The water molecules will form solvation complexes with the exposed ionic constituents of the minerals at the disrupted sites. Besides, the exchangeable ions (e.g. Na^+ or Mg^{2+}) on the surfaces of clay minerals or on metal hydroxides can solvate easily and diffuse away. But the framework ions closely bound in crystal lattices cannot be dislodged so easily. The solvation of such ions requires either the weakening of the bonds or an attack by a highly polarizing species, such as protons.

The calcium ions and to a lesser extent, the sulfate ions, enhance the plant nutrient status of the soil. Though the application of gypsum changes the pH by less than 0.3 log units only, it does bring down the concentration of Al in the soil solution and thereby indirectly improves the soil fertility.

2. 7   BIOGEOCHEMICAL PROCESSES IN SOILS

2.7.1   Exchangeable Ions in Soils

The soil humus contains a variety of functional groups. They include the carboxyl group (COOH) and phenolic hydroxyl group (aromatic ring OH) which can dissociate a proton and become negatively charged in the soil solution (see Table 2.8). Besides, there are surface functional groups which are molecular units that protrude from the solid surface into the soil solution. Functional groups on surfaces cannot be diluted infinitely, even in aqueous suspension.

"Adsorption is the net accumulation of matter at the interface between a solid phase and an aqueous solution phase" (Sposito, 1989, p.132). Adsorption generally involves a two-dimensional molecular arrangement (as against a three-dimensional molecular structure in a precipitate). A simple example is the adsorption of the monovalent cation, K^+, on the surface of the 2:1 layer silicate montmorillonite. Only fully solvated ions adsorbed on soils can be deemed to be *readily exchangeable ions*.

Ion-exchange Reactions

In the course of an ionic exchange reaction, an ionic species in a solid compound is replaced by another ionic species taken from an aqueous solution in contact with the solid:

$$CaCO_3 \text{ (s)} + Mg^{2+} \text{ (aq)} = MgCO_3 \text{ (s)} + Ca^{2+} \text{ (aq)} \qquad (2.6)$$

Chloride replacement by nitrate on a Ferralsol is a good example of an ion exchange reaction in soils. It may be noted that an ionic exchange reaction is not bond-specific (i.e., it is not confined to any particular kind of bonding, such as electrostatic, ionic or covalent).

The Cation Exchange Capacity (CEC) values (moles of charge per kilogram of adsorbent — $mol_c \ kg^{-1}$) for various soils are as follows (the data are based on measurements made for U.S. soils using NH_4^+ as the "index" cation at pH 7; Holmgren et al., quoted by Sposito, 1989, p. 171). As the soil orders are as per the US system, the corresponding soil grouping as per the FAO-Unesco-ISRIC system, 1988, are given in parentheses.

Alfisols (Luvisols): 0.12 ± 0.08; Aridisols (Calcisols, Gypsisols, Solonchaks): 0.16 ± 0.05, Entisols (Fluvisols, Regosols): 0.13 ± 0.06; Histosols: 1.4 ± 0.3; Inceptisols (Andosols, Cambisols and Gleysols): 0.19 ± 0.17; Mollisols (Chernozems, Phaeozems and Kastanozems):

0.22 ± 0.10; Oxisols (Ferralsols, Nitisols): 0.05 ± 0.03; Spodosols (Podzols): 0.11 ± 0.05; Ultisols (Acrisols): 0.06 ± 0.06; and Vertisols: 0.37 ± 0.08.

CEC values vary for different soil types, and in different layers of the same soil (say, Podzols). Generally, Acrisols, Ferralsols and Nitisols are characterized by markedly lower values than Histosols and Vertisols; and this is related to their organic matter and clay contents.

Figure 2.5. shows that the content of organic matter (organic Cg kg^{-1}) markedly influences the CEC values.

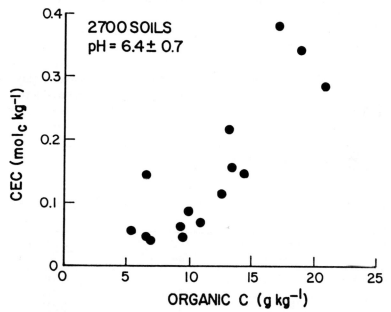

Fig. 2.5. Relationship between the soil CEC and the soil organic matter content (*source:* Sposito, 1989, p. 172) © Oxford University Press.

2.7.2 Soil Colloidal Phenomena

Colloids are small (0.01 to 10 μm diam.), virtually insoluble particles. In the case of soils, clay minerals and metal hydroxide particles constitute the colloids. These particles do not dissolve in water but remain in suspension. Colloids which do not undergo gravitational settling in periods of less than 24 hours, are said to be stable.

Soil erosion and clay illuviation are effected through the mechanism of stable suspensions of colloids. Trace metals, phosphate anions or pesticides may get adsorbed on the soil colloids, and move with them. Thus, colloids play a significant role in the particle and chemical transport in soils.

A colloidal suspension coagulates when it becomes unstable and undergoes gravitational settling. Coagulation processes control the formation of soil structure. *The critical coagulation concentration (ccc) is the* smallest concentration of electrolyte (in mol m^{-3}) at which the soil colloidal suspension undergoes rapid coagulation. *The Schukze-Hardy Rule*, which governs the coagulation process, may be stated as follows: "The critical coagulation concentration for a colloid suspended in an aqueous electrolyte solution is determined by the ions with a charge opposite in sign to that on the colloid and is proportional to an inverse power of the valence of the electron."

The stability of soil colloidal suspensions is affected by the following factors: electrolyte concentration, pH value, adsorption of small ions and adsorption of polymer ions.

Table 2.11 (*source*: Sposito, 1989, p. 201) gives the mean values and standard deviations of *ccc* for ions, whose absolute valence, $|Z|$, is equal to 1 or 2. In the case of metal oxides, the particles have a positive charge, and the coagulating ions have a negative charge. In the case of clays, the particles have a negative charge, and the coagulating ions have a positive charge. The ratio of *ccc* values for $|Z| = 2$ and $|Z| = 1$, is given in the fourth column of the Table. It may be noted that the theoretical value predicted for the ratio in the last column ($2^{-6} = 1/64 = 0.0156$) is of the same order as the ratio between the experimentally determined *ccc* values for $|Z| = 2$ and $|Z| = 1$.

Table 2.11 Critical coagulation concentrations (ccc) for colloidal suspensions of soil minerals

Soil mineral	ccc ($|Z| = 1$) (mol m^{-3})	($|Z| = 2$) (mol m^{-3})	$\dfrac{(|Z| = 2)}{ccc\,(|Z| = 1)}$
Al hydrous oxide	50 ± 9	0.5 ± 0.2	0.010
Fe hydrous oxide	11 ± 2	0.21 ± 0.01	0.019
Illite	48 ± 11	0.14 ± 0.02	0.003
Kaolinite	10 ± 4	0.3 ± 0.2	0.030
Montmorillonite	8 ± 6	0.12 ± 0.02	0.015

2.7.3 Cation and Anion Adsorption in Soils

Negative charges on soil colloids are responsible for cation exchange. The negative charges may be of permanent nature arising from the isomorphous substitution of a clay mineral constituent by an ion with a lower valency, or it may be pH-induced. Generally, increasing the soil pH, at least up to neutrality, tends to increase CEC. "Humic polymers in the soil organic matter fraction become negatively charged due to the dissociation of protons from carboxyl and phenolic groups" (Alloway and Ayres, 1993, p.38).

CEC is easily the most important contributory factor to fertility. CEC can be enhanced by the addition of organic matter and clays and where the soils are acid, raising the pH nearer to neutrality.

The ability of one cation to replace another cation is a function of its valency and the diameter of its hydrated form. Thus, Mg^{2+} can replace Cs^+ because of its higher valency, and Ca^{2+} can replace Mg^{2+} because of the greater size of the hydrated Mg^{2+} ion. The order of replaceability is as follows:

$$Li^+ = Na^+ > K^+ = NH_4^+ > Rb^+ > Cs^+ > Mg^{2+} > Ca^{2+} > Sr^{2+} = Ba^{2+} > La^{3+} = H^+ (Al^{3+}) > Th^{4+}.$$

Anions are adsorbed on positively charged soil colloids, such as hydrous oxides of Fe and Al.

2.8 BIOGEOCHEMICAL CYCLES

As already indicated, elements get recycled within and between the various components of the geoenvironment (lithosphere, hydrosphere, atmosphere, and biosphere). The recycling processes are energized by solar radiation (ultra-violet, visible and infrared), mechanical (kinetic and potential), chemical and thermal energy of the earth (part of which is derived from the decay of uranium, thorium, and [40]potassium). In most cases, the elements are released during the process of weathering, transported, and recombined in various ways. The transport mechanisms are indicated in Fig. 2.6.

The biogeochemical cycle is schematically shown in Fig. 2.7

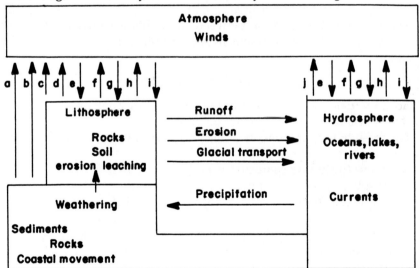

Fig. 2.6 Transport mechanisms
(*source:* Fergusson, 1990, p. 147, © Author)
(a) volcanic activity, (b) weathering, (c) weathering, (d) aerosol, (e) fallout (solid),
(f) outgassing, (g) gas absorption, (h) evaporation, (i) precipitation, (j) spray

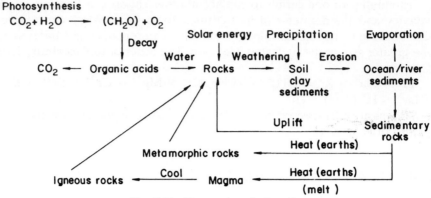

Fig. 2.7 Biogeochemical cycle
(*source:* Fergusson, 1990, p. 147 © Author).

2.8.1   Residence Times, Pools, and Fluxes

The dynamics of the environmental components is best understood through box models. At any given point of time, each box (reservoir) has a pool of an element. Each box receives an input and gives out an output. Where the rate of output of a reservoir is far more than that of the input, it constitutes a source (e. g. erosion on land). Where the rate of input into a reservoir is far more than that of the output, it constitutes a sink (e. g. sea-floor sediment). (Garrels et al., 1975)

The movement of material between the pools is called the flux. The residence time (t) of an element in a reservoir is determined from the equation,

$$t = m/(dm/dt) = m \, (dt/dm),$$

where m = mass of the element in the reservoir or pool, and dm/dt = rate of input (output) to the pool.

The mass of the components of the geoenvironment (in $\times 10^{18}$ g) are as follows:

Biosphere (including contained water)	8
Hydrosphere:	
Fresh water (liquid)	126
Ice	30,000
Oceans	1,420,000
Salt in oceans	49,000
Atmosphere	5,140
Crust (down to 17 km.)	24,000,000

The general composition of components of geoenvironment (in wt %) is summarized below:

Crust :   O (46.6), Si (27.7), Al (8.1), Fe (5.0), Ca (3.6), Na (2.8), K
               (2.6), and Mg (2.1)
Atmosphere: N (78), O (21), Ar (0.9), CO2 (0.3), Ne (0.002),
               Others (< 0.01)
Biosphere: H (49.8), C (24.9), O (24.9), N (0.27), Ca (0.07),
               K (0.04), Mg (0.03), Si (0.03), P (0.03),S (0.02).

Detailed chemical composition of the earth's crust and sediments, chemical constituents of sea water and fresh water, and of the troposphere, have been summarized by Archer et al. (1987, pp. 139-141).

Because of various uncertainties, the residence times of different elements in different pools are not known accurately. In some cases, the residence times can be determined experimentally by using stable isotope tracers.

As should be expected, there is an enormous variation in the residence times of different elements in the different components of the geoenvironment. The residence times for heavy elements (As, Cd, Hg, and Pb, etc.) given below, are indicative of the large ranges:

Atmosphere   : 7-90 days
Hydrosphere:
    Surface ocean waters : A few (say, 2) years
    Marine sediments   : $2\text{-}5 \times 10^8$ years
    Lakes   : 300-700 days
Lithosphere (soils)   : 200-3,000 years
Biosphere (humans)   : 8-500 days

The units in which the pools, concentrations, and fluxes in the atmosphere, lithosphere, and hydrosphere are given, are as follows:

Sphere	Concentrations	Pools	Fluxes
Atmosphere	$ng\ m^{-3}, \mu g\ m^{-3}$	kg, t (tonne)	$kg\ y^{-1}, t\ y^{-1}$
Lithosphere	$ng\ g^{-1}1, \mu g\ g^{-1}$	kg, t	$kg\ y^{-1}, t\ y^{-1}$
Hydrosphere	$ng\ g^{-1}, \mu g\ l^{-1}$	kg, t	$kg\ y^{-1}, t\ y^{-1}$

$1\ kg = 10^3\ g$, $1\ t\ (ton) = 10^3\ kg = 10^6\ g$, $1\ \mu g = 10^{-6}\ g$,
$1\ ng = 10^{-3}\ \mu g = 10^{-9}\ g$, $1\ pg = 10^{-12}\ g = 10^{-3}\ ng = 10^6\ \mu g$.
$1\ ppm$ (by wt.) $= 1\ \mu g\ g^{-1} = 1\ mg\ kg^{-1} = 1g\ t^{-1}$
$1\ ppb$ (by wt.) $= 1\ ng\ g^{-1} = 1\ \mu g\ kg^{-1} = 1\ mg\ t^{-1}$

If the size of the pool, the concentration level of the element in the pool at present, and the fluxes are known, it is possible to predict the concentration levels at a given time in future, assuming that there is no temporal change in the parameters.

The box model whereby the movement of an element from one box to another is traced, is an oversimplification. There are two complications: (i) the speciation of an element could change depending upon the environment; for instance, mercury gets methylated by bacteria in the sediment, and methyl mercury is far more toxic than the metallic mercury; and (ii) an element exists in a box along with other elements, with which it may interact and form complexes, with different properties and mobilities.

2.8.2 Chemical and Isotopic Techniques of Tracing

Chemical and isotopic techniques are used to understand and quantify the dynamics of the processes of the geoenvironment.

Radiogenic isotopes (Pb isotopes) have been used by Patterson (1980) to trace the enhancement of lead in human environment with industrialization. The isotopes most used for environmental studies are the stable isotopes. Light elements with mass numbers of less than 40 (such as, H, C, O, and S) undergo fractionation due to physical, chemical, and biological factors. H, C, O, and S are important elements in the biosphere and hydrosphere, and to some extent in the lithosphere, and hence their importance in environmental studies.

The extent of fractionation in stable isotopes is quite small and it is customary to express the enrichment/depletion of the heavier of the isotopic pair in relation to a standard, in terms of parts per mille, i.e., parts per thousand (‰).

Element	Isotope pair	Fractionation	Standard
H	^1H, ^2D	δD	SMOW
C	^{12}C, ^{13}C	$\delta^{13}C$	PDB
O	^{18}O, ^{16}O	$\delta^{18}O$	SMOW, PDB
S	^{32}S, ^{34}S	$\delta^{34}S$	Troilite

SMOW = Standard Mean Oceanic Water, PDB = Pee Dee Formation, Belemnite; Troilite (FeS) phase of Canyon Diablo meteorite.

For instance, $\delta^{13}C$ (‰) is calculated as follows:

$$\delta^{13}C(‰) = \frac{(^{13}C/^{12}C)\,sample}{(^{13}C/^{12}C)\,standard} \times 1000$$

Terrestrial soils generally have $\delta^{18}O$ values ranging from +15 to +25‰, and δD values from −30 to −100‰. Soils rich in gibbsite or other Al-Fe oxides can be distinguished from clay-rich soils by being isotopically lighter in ^{18}O. Kaolinite and montmorillonite are about 27‰ richer in ^{18}O than the coexisting water, whereas gibbsite is 18‰ richer in ^{18}O. On the other hand, all the clay minerals and hydroxides are depleted in deuterium relative to the co-existing water: kaolinite and montmorillonite: about 30‰, and gibbsite: about 15‰. The isotopic composition of meteoric waters in the geological past can be delineated on the basis of $\delta^{18}O$ and δD of ancient kaolinites.

$^{18}O/^{16}O$ of detrital minerals in a soil can be used to delineate the provenance and mode of origin of the soil. For instance, detrital quartz is resistant to weathering and retains the original $^{18}O/^{16}O$ of the parent rock.

2.8.3 Speciation

"Species" has been defined as "molecular representation of a specific form of an element". It is now widely recognized that the environmental pathways and toxicity of an element are dependent on the speciation of the element concerned. For instance, hexavalent chromium is more toxic than the trivalent form. Methyl mercury, formed in aquatic environments due to bacteria, is far more toxic than metallic mercury. The most important pathway of mercury to man is through methyl mercury in fish. Aluminum gets mobilized in soils and waters under the conditions of acid rain.

CHAPTER

3

Soil Nutrients

A woman is like a tea bag — you know her strength only when she is in hot water

—Nancy Reagan.

3.1 PLANT GROWTH

It is not possible to state precisely what is meant by a fertile soil, since the requirements of plants differ widely. For instance, plants such as barley, maize and sugarcane require well-aerated soils, whereas anaerobic conditions are required in the early part of the cycle of cultivation of paddy. A soil may have inherent fertility, arising out of its mineralogy, humus content and ability to hold moisture. Fertility may be induced in the soil by the addition of suitable fertilizers.

The following factors affect plant growth: root-room and root-hold, aeration, moisture, temperature, essential elements, pH, and stable sites.

The chemical elements in the soil that are essential to plant growth are generally divided into three categories:

(i) *Micronutrients*, which are absorbed in small quantities: B, Cl, Mn, Fe, Cu, Zn, Mo, and Co; (ii) *Secondary nutrients:* Mg, S, and Ca; and (iii) *Macronutrients:* H, C, N, O, P, and K.

Plants need the essential elements to form plant tissues and to act as catalysts and intermediates in metabolic processes. The roles of individual elements are summarized as follows (FitzPatrick, 1993, pp.138-140):

C, H, O: major constituents of plant tissue; derived from atmosphere and water.

N: derived from atmosphere and dead tissues. It is transformed by soil bacteria into ammonia and nitrate which are taken up by plant roots. Nitrogen is a constituent of chlorophyll, and proteins, many of which serve as enzymes.

P : constituent of every living cell; occurs in protoplasm.

K: important in cell biotic processes.

Ca: Calcium pectate constitutes part of the cell wall. Ca is necessary for the growth of tips of stems and roots in most plants.

Mg: active in the enzyme system; forms part of the chlorophyll.

S: It plays an important role in the formation of some amino acids and oils.

Fe & Mn: They have a role in the enzyme system and are necessary for the synthesis of chlorophyll.

B: has a role in calcium utilization and development of actively growing parts of the plant.

Cu & Zn: form part of the enzyme systems and growth-promoting substances.

Mo: required for the reduction of nitrate in plants.

Cl: regulates osmotic pressure and cation balance in plants.

Co: plays a role in nitrogen fixation in leguminous plants.

It may be noted that with the exception of molybdenum, all the nutrient elements are "light elements" with atomic numbers of less than 30, and with relatively small ionic radii.

The concept of Ionic Potential (IP) is useful in understanding the behavior of elements in soil solutions. Ionic potential is obtained by dividing the Ionic Charge (Z) by the Ionic Radius (r, expressed in nanometers or 10^{-9} m) of an ion. Thus, the IP of Ca^{2+} with Z = 2, and ionic radius of 0.106 nm, is 19 nm^{-1}.

In the soil solution, elements with IP > 100 nm^{-1} (such as, B,C,N,P,S, and Mo) exist as oxyanions, whereas elements with IP < 30 nm^{-1} (such as, Mg, K, Ca, Mn, Fe, Cu, and Zn) exist as solvated cations. Elements for which 30 < IP < 85 nm^{-1} are potentially phytotoxic in dissolved form, e.g. Al (III), Ni (III), Hg (III), Cd (III), etc. It is possible that during the process of evolution plants did not find it necessary to develop detoxifying mechanisms to protect against these insoluble elements. Plants are now obliged to live with these toxic elements that have been introduced into the environment due to anthropogenic activities — and therein lies the problem.

The biological significance of these attributes of essential elements is best understood by an examination of the *Banin — Navrot plot* (Fig. 3.1), which is a log-log graph of the Biological Enrichmen Factor (EF_B) and Ionic Potential (IP).

$$EF_B = \frac{\text{Element concentration in organisms (mg kg}^{-1})}{\text{Element concentration in crustal rock (mg kg}^{-1})} \qquad (3.1)$$

$$IP = \frac{\text{Valence of element free cation}}{\text{Radius of element free cation (nm)}} \qquad (3.2)$$

It may be noted from Fig. 3.1 that the Banin-Navrot plots for higher plants and animals are remarkably similar. They also happen to bear close similarity to Banin-Navrot plots for soil microflora (bacteria and fungi).

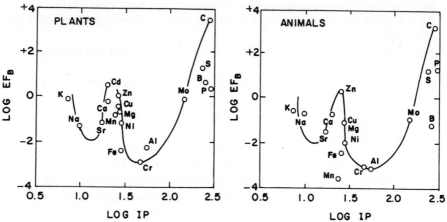

Fig. 3.1   Banin-Navrot plots for terrestrial plants and animals
(*source:* Banin and Navrot, 1985 © Science, USA)

An examination of Fig. 3.1 indicates that the Biological Enrichment Factor increases in the following order: hydrolyzed cations << solvated cations < oxyanion-forming ions. Obviously, *solubility* and consequent mobility of an ion are the determining factors for essentiality. In other words, an element not capable of solubilization could in no way become an essential/nutrient element.

3.2   PHYTOTOXICITY

It is difficult to prescribe the "optimal" solution concentrations of essential elements needed for plant nutrition because two or more elements may act synergistically, accentuating their combined effect, or antagonistically, suppressing the effect in part. For instance, zinc can moderate the toxicity of cadmium. At sufficiently high concentrations, even micronutrient elements such as B and Cl may become phytotoxic.

Plants have low concentration levels of hydrolyzate elements (such as, aluminum). Though Al is an abundant element in crustal rocks and soils, its intake by plants is low because of its low solubility. In the course of geological time, plants evolved under conditions of such a low availability of aluminum. Under conditions induced by anthropogenic activities, such as acid rain and consequent acid pH, some phytotoxic elements become more soluble and more available. Similarly, plants are unable to cope with high concentrations of copper (IP: 0.27 nm^{-1}), as they evolved under conditions of low levels of copper availability and did not develop mechanisms to detoxify the element when encountered in high concentrations.

Some of the mechanisms through which phytotoxicity arises are: (i) binding by bioligands can effectively displace an essential element (e.g. Ca) by a toxic element (e.g. Cd); (ii) biocomplexation of a toxic metal by a functional group in a biomolecule may render the group incapable of reacting further; and (iii) the biochemical function of a biomolecule may be affected because of interaction with a toxic metal. Evidently, complex formation plays a critical role in phytotoxicity, and it follows that the stronger the tendency of a metal to form complexes, the greater the possibility of its being phytotoxic.

An understanding of the mechanisms of phytoxicity enables us to design strategies for the mitigation of the deleterious consequences of anthropogenic phytotoxins.

The toxicity sequence for each class of plants indicates the *order* of concentration of metal required to produce a particular degree of toxicity. It follows that the more toxic an element, the less the amount required to produce a particular toxic effect. *Per contra*, a less toxic element will be required in larger quantities to produce the same level of toxicity. For each class of plants, the order of metals from left to right reflects an increasing concentration of the metal (in mol m^{-3}). It may be noted that the toxicity sequences in Table 4.1 (*source:* data of Nieboer and Richardson, quoted by Sposito, 1989, p. 250) are remarkably similar. Mercury leads the sequence in most cases. Its toxicity arises because of its strong tendency to form organo-metallic complexes.

Table 3.1   Metal toxicity sequences for various organisms

Organisms	Toxicity sequence*
Algae	Hg >Cu >Cd >Fe >Cr >Zn >Co >Mn
Flowering plants	Hg >Pb >Cu >Cd >Cr >Ni >Zn
Fungi	Ag > Hg >Cu > Cd > Cr > Ni >Pb > Co > Zn > Fe
Phytoplankton (fresh water)	Hg > Cu > Cd > Zn > Pb

* The speciation involved is as follows: Hg = Hg(II), Fe = Fe(II), Cr = Cr (III), Co = Co (II), Mn = Mn (II), Pb = Pb (II)

3.3   GEOCHEMICAL AND BIOAVAILABILITY

3.3.1   Trace Metal Cycling in Soils

Trace elements (TE) occur in soils in the following forms (in the order of increasing availability to plants): (i) dispersed in silicate crystal lattices, e.g. feldspars, clays; (ii) in the form of their own compounds, e.g. PbS, $ZnCO_3$; (iii) dispersed in hydrated oxides, e.g. Fe and Mn oxides; and (iv) held on surfaces by ion-exchange or physical adsorption, e.g. clays, organics. The soluble forms of TE are the hydrated ions, and complex ions with OH^-; Cl^-, and organic acids as ligands (Fig. 3.2).

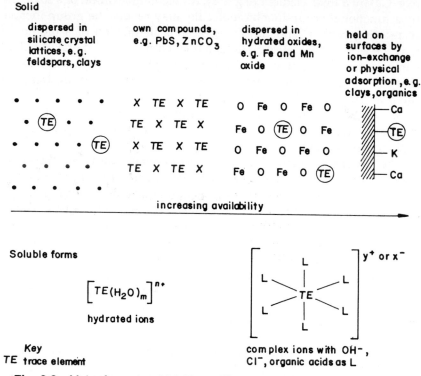

Fig. 3.2 Major forms in which Trace Elements (TE) are found in soils
(*source:* Peter O'Neill, 1985, p. 189 © George Allen & Unwin)

Trace-metal cycling plays an important role in the soil — plant relationship. The phytoavailability of metals control their supply to plants and therefore their flux to the food chain. The metals that bioaccumulate in the surface soil horizons get partly immobilized in the surface soil horizons, thereby preventing their transfer to the hydrocycle. The speciation and localization of metals in the soils, depend upon (i) the nature of the mineral inheritance from the parent material, (ii) the kind of complexes that develop during the soil processes, and (iii) the types of compounds formed as a consequence of anthropogenic impact. Thus, the behavior and phytoavailability of metals in soils depend upon whether the soil is lithogenic, pedogenic, or anthropogenic. The binding mechanisms of trace metals in soil are related to the mechanical composition of the soils, quality and quantity of organic matter, reactions in the soils, redox, and drainage conditions. Thus, an understanding of the properties and behavior of trace metals is necessary to predict their fate in the soil envionment (Alina Kabata-Pendias, 1991, Abstract of the Third Symp. on Environmental Geochemistry & Health, Uppsala, Sept. 91).

Trace elements in some form are more available to plants than other forms. Plants can readily take up TE when they are in the form of soluble ions or complex ions, provided that the plant root has no specific exclusion mechanisms. The next group are adsorbed ions. They become more available when the conditions become more acid and hydrogen ions displace the cations. Any change in the soil environment that has the effect of increasing the solubility of TE promotes increased availability, such as reducing conditions (as in waterlogged soils), increased acidity that promotes the solubility of hydrated oxide, etc.

A high proportion of a particular element may be bound up in a soil mineral in an unavailable form, e.g. phosphorus in the rock phosphate, apatite. In order to make the phosphorus in apatite available to the plant, apatite fertilizer needs to be crushed and acidulated before application.

The mode of association of trace elements with various components of the soil can be estimated by means of different extracting solutions:

Ammonium acetate solutions extract easily exchangeable metals (principle: NH_4^+ ions will occupy the cation-exchangeable sites, releasing the metals previously bound to these sites);

EDTA extracts more tightly held ions associated with organic matter.

3.3.2 Bioavailability

The quantity of trace element that reaches man through food is determined by (i) geochemical availability (depending upon the leachability of the element and the mode of its distribution and availability in the rock or soil), and (ii) bioavailability (the fraction of the element in the food of plant and animal origin available to man) (Fig. 3.3).

The geochemical pathways by which nutrient or toxic elements reach man from various sources, such as soil, are shown schematically in Fig. 3.4.

A chemical element becomes bioavailable under the following conditions: (i) it should be present as a free ion species or it should be capable of being transformed readily to the free ion species (the significance of this condition arises from the fact that elements move in the soil environment through diffusion and convective flow in water); (ii) it should be capable of moving to the plant roots on a time scale that is relevant to plant growth and development (an element that moves so

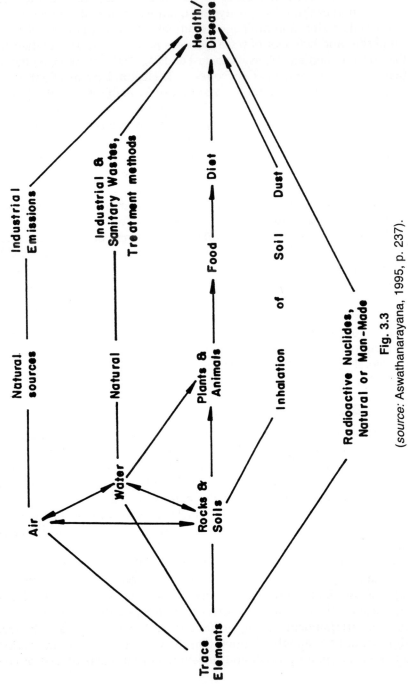

Fig. 3.3

(*source:* Aswathanarayana, 1995, p. 237).

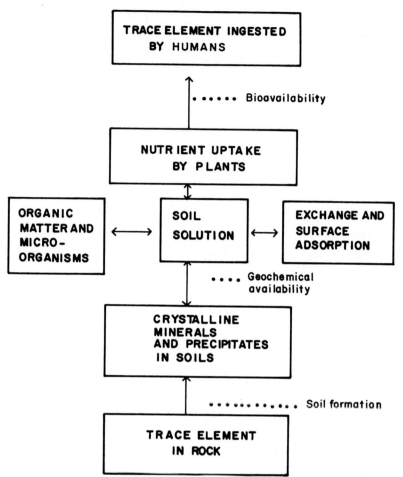

Fig. 3.4 Geochemical and bio-availability of elements
(*source:* Aswathanarayana, 1995, p. 237)

slowly that it cannot reach the plant roots during the period of growth of the plant is useless to the plant for all practical purposes); and (iii) it should be capable of affecting the life cycle of the plant when once absorbed by the root (an essential element promotes growth and development and a toxic element produces phytotoxicity). In the case of an element which has no effect on the plant, good or bad, it makes no difference whether the element concerned reaches the plant or not.

Which element and in what quantities will be able to reach the plant are determined by the competition among the plant-root system, the soil solution and the soil solid phases (Fig. 3.5).

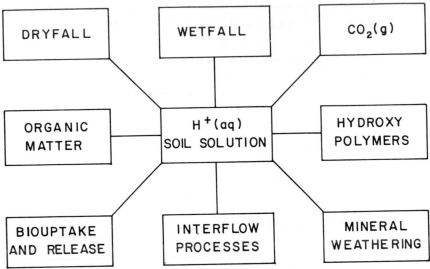

Fig. 3.5 Schematic illustration of the competition for free aqueous ions among the functional groups in the soil
(*source:* Sposito, 1989, p. 251, © Oxford University Press).

Several elements which can exist as ionic solutes compete to get into the soil solution. This involves complexation reactions with both inorganic and organic ligands or with metal ions.

An examination of Table 2.10 dealing with the speciation of various elements in acid and alkali soils (source: Sposito, 1989, p. 69), suggests that the complexation of an ion becomes more pronounced with increasing ionic potential (Ca^{2+} with IP of 17.9 nm^{-1} has a more pronounced tendency for complexation than Na^+ with IP of 9.8 nm^{-1}; similar consideration applies to Cu^{2+} with IP of 35.1 nm^{-1} vis-à-vis Ca^{2+} with IP of 17.9 nm^{-1}). The relevance of the complexation tendency arises from the fact that the complexing ligands must be shed before the metal concerned can be absorbed by the plant roots. This does not apply to the "scavenger compounds", such as the amino acid siderophores that complex Fe and Cu.

The inorganic-organic solid surfaces compete for free-ion forms of chemical elements. Inorganic solids, such as clay minerals, phosphates and metal oxides, which form rapidly but weather slowly, compete effectively for free ions. The deficiency of potassium arising from the adsorption of K^+ on 2:1 layer-type clay minerals, deficiency of copper arising from the adsorption of Cu^{2+} on soil organic matter, reduction in Al toxicity caused by the specific adsorption of Al-hydroxy species, are some of the consequences of the competition.

The soil organic matter is the chief repository for C, N, P, and S. Where the soil organic matter has undergone humification, its ability to retain the essential elements gets strengthened considerably. This explains the greater effectiveness of inorganic fertilizers when applied in combination with organic manure.

Surface complexation and diffuse-ion swarm association are some of the mechanisms for the adsorption of some important nonpolymeric anions in the soil solution, such as B $(OH)_4^-$, CO_3^{2-}, HCO_3^-, H_3SiO_4, PO_4^{3-}, HPO_4^{2-}, MoO_4^{2-}, etc.

There is convincing field and laboratory evidence to suggest that plant absorption is critically dependent upon the free-ion species. Positive correlation has been reported between metal uptake or concentration in plants and the thermodynamic activity of the metal (Fig. 3.6, for Fe and Mn in the case of rice).

3.4 KINETICS OF NUTRIENT UPTAKE

The processes of nutrient uptake take place in the *rhizosphere*, which is the local soil environment that is influenced significantly by plant roots. Convective flow and diffusion are the two principal mechanisms for the movement of elements (nutrient or otherwise).

"Convective flow of a nutrient occurs only when a plant transpires and its roots absorb water from the rhizosphere" (Sposito, 1989, p. 253). The nutrient uptake flux is the amount of nutrient absorbed per unit cross-sectional area of root per unit time. It is numerically equal to to the product of nutrient concentration and the water absorption rate. It is of the order of $\times 10$ μmol cm^{-2} d^{-1}. The distance over which the convective flow to the root occurs is of the order of a few mm (typically 3 mm). Convective nutrient uptake takes place in a matter of hours. This time scale is consistent with the fact that water absorption takes place during the daylight only.

Convective uptake is particularly important in the case of relatively mobile nutrients, such as NO_3^-, $H_3BO_3^\circ$, and Ca^{2+}. It may be the dominant mechanism operating in fertile soils.

The uptake of nutrients through diffusion is a much slower process, but it happens to be the dominant process in the case of relatively immobile elements like P and Cu, particularly when their concentrations in soil solution are low.

The laws of diffusion of ions across a film can be adapted to study the diffusive nutrient uptake. According to *Fick's law of diffusion*,

$$J = \frac{D}{\delta}(c - c')$$

(3.3)

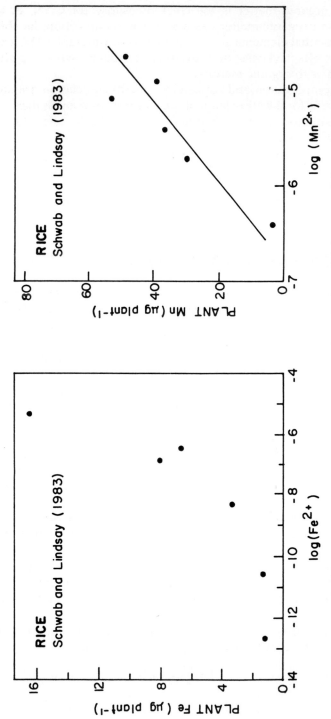

Fig. 3.6 Relationship between the metal uptake by the rice plant and the aqueous free metal ion activity
(*source:* Schwab and Lindsay, 1983, © Soil Soc. Amer).

where J is the rate of diffusion from bulk solution to exchanger surface per unit area of exchanger surface (mol $m^{-2} s^{-1}$), D the diffusion coefficient of the ion ($m^2 s^{-1}$), δ thickness of the film, c and c' are respectively the concentrations of the ion in the bulk solution and at the film-exchanger interface.

Nutrient species such as NO_3^-, SO_4^{2-}, $H_3BO_3^-$, whose time scale for nutrient diffusion (δ) is of the order of a day, get into the plants readily. The time scale is large for nutrients (such as, $H_2PO_4^-$) which have an affinity for soil solid phases. For nutrients which need thousands of days to be available to the plant, deficiencies are likely to arise, particularly in the absence of significant convective uptake or direct uptake by the plant root itself.

3.5 SOIL pE AND pH EFFECTS

The chemistry of the soils is strongly dependent on soil pE and pH (pE = Eh/0.059).

The pE – pH diagram (Fig. 3.7) shows the domain in which the soil microorganisms function, and the electron and proton activity levels commonly observed in the soils. At pH 7, *Oxic* soils have pE > +7, *suboxic* soils have pE in the range of +2 to +7, and *anoxic* soils have pE of less than +2. pE and pH have an inverse relationship. If pE increases, pH must decrease. This inverse relationship between pE and pH is manifested by the slanting lines separating the domains of oxic, suboxic and anoxic soils. Thus, a higher pE is needed for oxic conditions in acid soils than in alkaline soils.

C, N, O, S, Mn, and Fe are the elements most affected by soil redox reactions. When a soil is contaminated by anthropogenic activities, other elements, such as As, Se, Cr, Hg, and Pb, enter the picture. There are well-defined pE limits for the functioning of microorganisms in the soil. For instance, aerobic microorganisms do not function below pE 5. Denitrifying bacteria function in the pE range of +10 to 0. Sulfate-reducing bacteria cannot function at pE values above + 2, and so on. Thus, pE – pH diagrams help us to understand the stability conditions of chemical and microbial species in the soil.

Soil pH could cause direct toxicity for plant roots (for instance, the Al, Fe, and Mn toxicity in acid soils causes the loss of root membrane integrity), and to microorganisms. It indirectly affects bioavailability as protonation could enhance the production of free metal cations in the soil solution.

pE has both *direct* and *indirect effects* on the bioavailability of plant nutrient elements.

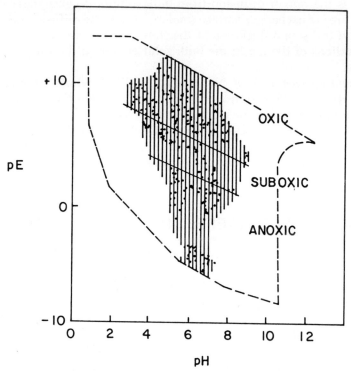

Fig. 3.7 A pE–pH diagram showing the area accessible to micro-organisms (dashed perimeter) and that observed in the soil (shaded area) (*source:* Sposito, 1989, p. 107 © Oxford Univ. Press).

The mineralization of soil organic matter and the precipitation of hydroxy solids constitute the most important direct results of increasing soil pE (Table 3.2; *source:* Sposito, 1989, p. 257). This process enhances the concentration of free ion species of N, P and S in the soil solution, and increases their bioavailability. At the same time, the process leads to the immobilization of Fe and Mn, thus decreasing their bioavailability. In the pE range of 3 – 9 (which corresponds to the transition interval for oxygen-nitrogen respiration by soil microorganisms), the bioavailability of nitrogen is profoundly influenced by the microbially catalyzed processes of nitrification, nitrogen fixation, and ammonia volatilization.

Adsorption phenomena constitute the most significant indirect effects of decreasing soil pE (vide Table 3.3; *source:* Sposito, 1989, p. 258). Under suboxic to anoxic conditions (pE < 3), the Fe and Mn hydrous oxide adsorbents get destabilized, thus facilitating the increase in the free-ion concentrations of anionic nutrients. In general, low pE enhances the

Table 3.2 Direct effects of soil pE on the bioavailability of nutrient elements, through the mineralization of soil organic matter and the precipitation of hydroxy solids (g = gas; L = liquid; aq = aqueous)

Nutrient element	Effects of pE on the bioavailability of the nutrient element
Carbon	Conversion of organic forms to HCO_3 and CO_2 (g); CO_2 (aq) to CH_4 (g)
Nitrogen	Conversion of organic forms to NH_4 ; NH_4 to NO_2 to NO_3; NO_3 to N_2 (g) and N_2O (g); NH_4 and NH_3 (g); N_2 (g) to organic forms
Oxygen	Conversion to H_2O (L) and other oxide forms
Phosphorus	Conversion of organic forms to H_2PO_4 and HPO_4
Sulfur	Conversion to organic forms of SO_4
Manganese	Precipitation of Mn (IV) hydroxy solids
Iron	Precipitation of Fe (III) hydroxy solids

Table 3.3 Indirect effects of soil pE on the bioavailability of nutrient elements through the control of adsorption processes

Element	Effect of PE on bioavailability
B, C, P, S, Mo, Cu, Zn	Dissolution of Fe (III) and Mn (IV) hydroxy solid adsorbents (specific adsorption)
N, S, Cl	Dissolution of Fe (III) and Mn (IV) hydroxy solid adsorbents (nonspecific adsorption)
Mn, Fe, Cu, Zn	Stabilization of polymeric, metal-complexing organic matter; production of inorganic sulfides
K, Mg, Ca	Competition with NH^{4+}, Mn^{2+}, and Fe^{2+} for cation exchange sites
H	Conversion to H_2O (L) via the reduction of oxide forms of N, S, Mn and Fe

bioavailability of anionic nutrient elements (particularly, B, inorganic C, P, and Mo). It does not seem to affect the micronutrient metals.

Low pE conditions promote the concentration of some nutrient cations, such as NH_4^+, Mn^{2+}, and Fe^{2+}. These will compete with other nutrient cations, such as K^+, Mg^{2+}, and Ca^{2+} which are readily exchangeable on soil particle surfaces. Ultimately, which nutrient ions will reach the plant will depend on the consequences of the competition.

3.6 SOIL NUTRIENTS

3.6.1 Nutrient Elements in the Common Fertilizers

Cultivation of land to grow crops appears to have started about 10,000 years ago, in the fertile Indo-Gangetic alluvium, deltas of the Nile, Euphrates-Tigris, etc. The use of manure for fertilizing the soil is probably as old as agriculture itself. The Romans had even a particular God, Sterculius, to preside over the protection of fertility of the soil.

The use of commercial fertilizers is relatively of recent origin. The use of superphosphate, saltpeter and guano as fertilizers started around 1840. The consumption of fertilizers increased rapidly; for instance, the consumption of fertilizers in the world in 1981/82 was 100 times more than it was at the beginning of the century, and 10 times more than it was in the late 30s. Fertilizer consumption (in kg of plant nutrients per hectare of arable land) varies enormously among countries. It ranges from a low of 0.8 kg ha^{-1} in Mozambique to a high of 723 kg ha^{-1} in Ireland. Though China (262 kg ha^{-1}) and India (69 kg ha^{-1}) have relatively higher figures, the figure is less than 40 kg ha^{-1} for most of the low-income countries. For higher income countries, the average is about 120 kg ha^{-1} (all figures from the World Development Report, 1992, pp. 224 and 225, of the World Bank).

The common fertilizers and the nutrient elements they provide are as follows (FitzPatrick, 1993, p.142):

Sulfate of ammonia: 20.5% N; Urea: 45% N; Ammonium nitrate: 35% N; Rock phosphate: 11 – 15% P; Superphosphate: 7 – 8% P; Basic slag: 2 – 8% P; Bone meal: 7 – 13% P; Potassium chloride: 39 – 42% K.

The increase in harvest due to the use of fertilizers varies greatly with the nature of the soil, agroclimatic conditions, crop management, etc., but a rough idea of the expectable increment in harvest due to the application of fertilizers, other things being equal, is nevertheless useful. It has been reported that the application of 1 kg of nitrogen leads to a harvest increment of 16 feed units, 1 kg of phosphorus 10 feed units, and 1 kg of potassium 4.5 feed units (one feed unit corresponds to the energy of 1 kg of barley). It has even been claimed that half of the world's population is fed by the extra yield produced by the use of commercial fertilizers.

The use of fertilizers is not an unmixed blessing. For instance, due to competition, excessive application of potassium leads to the deficiency of magnesium, an essential element. Copper poisoning of sheep has been reported in situations where the fodder has low molybdenum content. Some nitrogen compounds in the plant have a poisonous effect on animals. Nitrogen fertilizers cause pollution of groundwater. In the province of Haryana in India, groundwater has a nitrate content of 114-1,800 mg L^{-1} (as against the permissible content of 45 mg L^{-1}). Groundwater may also be contaminated by sulfate and chloride from fertilizers.

3.6.2 Soil nutrient Management

Soils continually receive plant wastes. If the soil has a basic deficiency in some element (for instance, selenium deficiency in Finnish soils), the application of farmyard manure alone cannot correct the deficiency. The deficient element has to be brought from elsewhere and applied. Generally, if a soil is deficient in a particular nutrient element, this will be reflected in the vegetation. Where the deficiency is identifiable, it can be corrected by appropriate fertilization, say, of magnesium, phosphorus, calcium or copper, as necessary.

Disturbed nutrient balances in the soils manifest themselves in the form of diseases of animals. Animals fed with crops from rice-berseem rotation grown on recently reclaimed salt-affected soils, died of "molybdenosis", caused by Mo-Cu imbalance. Mo-induced Cu deficiency can be cured with a small dose of copper sulfate.

Deficiencies and toxicities of some elements in the soils arise from the nature of the soils, management practices, and cropping systems. The commonly noted problems are the field-scale deficiencies with respect to Zn, Mn, Fe, and S, and toxicity of Se. Alkaline soils have severe deficiency of not only zinc but also manganese. The phenomena of micronutrient stresses, and synergism and antagonism among the nutrients, are now well recognized. For instance, phosphorus can induce Zn deficiency and molybdenum can induce copper deficiency.

Zinc

The dwarf Mexican wheat varieties which have an average yield of $5 - 6$ t ha^{-1} (as against about 1.5 t ha^{-1} for the traditional varieties) are highly responsive to fertilizer use. Zinc can be applied in the form of zinc sulfate, chloride, carbonate or oxide of zinc, zincated urea, zincated superphosphate, and chelated forms. The recommended dose is $5 - 10$ kg ha^{-1} of zinc sulfate twice, after 2 or 3 cropping sequences. Foliar sprays of zinc sulfate are suitable for horticultural tree crops and are compatible with pesticide sprays.

The solubility of zinc is profoundly influenced by soil pH, but not by the concentration level of sodium. When zinc fertilization is applied regularly to a rice-wheat sequence for three years, the build-up of zinc takes place in the following forms: amorphous zinc > crystalline Zn > complexed Zn > residual mineral Zn > exchangeable Zn. The Zn uptake in plants is largely dependent upon the exchangeable and amorphous zinc.

Generally, zinc is best applied through broadcasting and incorporation or placement in bands, rather than to foliage, seeds or roots of seedlings. The rate of zinc application varies with the soils — from 5.5 kg ha^{-1} in the case of sandy alkaline soils, 11 kg ha^{-1} in the case of calcareous and/or moderately alkaline soils, to 22 kg ha^{-1} in the case of floodplain and highly deteriorated alkaline soils. These ranges of zinc application take care of different crop rotations with a repeat zinc application after the fourth crop. In the case of sodic soils, deficiency of calcium and/or excess of sodium need to be ameliorated simultaneously with the amendment of zinc deficiency (Takkar and Bansal, 1987).

DTPA-extractable zinc in the soil is a good indicator of zinc availability to the plant. Zinc deficiency in wheat is indicated if the Zn content is less than 20 ppm. Zinc deficiency is graded as follows: less than 8 ppm = severe deficiency; 8–12 ppm = moderate deficiency, and 12–20 ppm = marginal deficiency. DTPA-extractable Mn level of 2.1 ppm in the soils is the critical value indicator of Mn deficiency. Through nutrient indexing of specific crops and soils, it is possible to take timely ameliorative action.

The commonly used micronutrient carriers are: manganese sulfate for Mn, ferric sulfate for Fe, ammonium sulfate for N, and gypsum for S.

Experience has shown that if legumes precede cereals in such situations, the fertilizer N application to the cereal crop could be reduced by 25–30 kg N ha^{-1}.

3.7   PHOSPHATIC FERTILIZERS

The complex physicochemical processes taking place in the soil control the nature and extent of the flow of nutrients to the plant roots. In the case of phosphorus, the behavior of P from a fertilizer source differs from that of equilibrium P in the soil. It is possible to increase the P use efficiency of the plants by optimizing those soil factors that promote P diffusion.

Phosphorus in strongly alkaline calcareous soils is present largely in the form of Ca P (54%), and residual inorganic P forms (28%). Leaching losses of P in alkaline soils can be considerably reduced if the gypsum requirement is mixed in the target zone (Gupta and Abrol, 1990).

High Fe content in the growth medium reduces the rate of entry of P into the roots. The P/Zn and P/Fe ratios are better indicators of the nutrient deficiency status of a soil than the absolute concentration levels of these elements.

The desirable minimum water solubility of the phosphorus fraction of the nitric phosphate fertilizer suitable for various soils and crops, has been found to be 50%. Application of nitric phosphates with 70% or more water-soluble sources led to the highest crop yields. The efficiency of low water-soluble phosphate fertilizer improved with time with regard to paddy-wheat cropping sequence. Acidulation through incorporation of pyrites, or mixing with farmyard manure and straw increased the effectiveness of phosphate fertilizer.

Phosphate rock is the principal raw material for the production of industrial phosphatic fertilizers. The principal phosphorus mineral is apatite, $Ca_5 (PO_4)_3 (F,Cl,OH)$. Apatite in magmatic and metamorphic rocks is crystalline fluorapatite or chlorapatite — $Ca_5 (PO_4)_3 (F,Cl)$, with a P_2O_5 content in the range of 42.3–41.0%. In sedimentary phosphates (phosphorites), phosphate may occur in the form of cement in sandstones, or as oolites and concretions. Phosphorites should contain at least 20% P_2O_5 to be commercially usable. In some countries, phosphorites with about 5% P_2O_5 are crushed and spread directly. Where phosphorus exists in a tightly bound form in the crystal lattice of igneous apatite, recourse has to be taken to crushing and acidulating the phosphatic rock before direct application. This procedure is necessary in order to make phosphorus available to plants.

The world production of phosphate rock is of the order of 200 million tons per annum. The USA, USSR, Morocco, China, Tunisia, Jordan, Brazil, Israel, South Africa, Togo, Nauru, Senegal, Syria, etc. are the important producers.

Raw sedimentary phosphate (phosphorite) invariably contains some amount of cadmium, which ends up in the industrial fertilizers. When such industrial fertilizers are applied, cadmium is taken up by the plants. Studies made in Sweden show that old cultivated soils tend to contain more Cd (0.057 mg kg^{-1}) than newly cleared land (0.044 mg kg^{-1}). Cadmium uptake by plants appears to be controlled by soil pH (higher Cd, at lower pH), amounts of Cd applied, and plant species.

Some phosphorites have very high uranium contents. For instance, the phosphorite of Minjingu, Tanzania, has 200 ppm of uranium and 0.57% of potassium. The gross radiation dosage at Minjingu has been estimated to be at least 150 μR h^{-1} or equivalent to 25 milligray y^{-1} (Aswathanarayana, 1988). Direct application of crushed phosphate (with about 30 % P_2O_5, and with 86% of the grains less than 0.25 mm) at the rate of $300 - 600$ kg ha^{-1}, has become popular with the farmers in Tanzania for the following reasons: (i) crushed phosphorite costs only one-third of the commercial superphosphate, and (ii) because of its high CaO content (about 40%), it makes a good liming agent and reduces the soil acidity (from 5.4 to 6.5). The direct application of Minjingu phosphorite at the rate of 500 kg ha^{-1} would involve the addition of about 6.5 g of P and 10 mg of U per m^2. The application of bentonite (which is fortunately available nearby) at the rate of 20 t ha^{-1} once in 7 to 10 years, markedly reduces the loss of nutrient elements through the leaching of P (and in the present case, U and F as well) by rain water, while concomitantly preventing the pollution of stream waters by U and F (Aswathanarayana, 1988).

3.8 SEWAGE SLUDGE

Sewage sludge is used in agriculture as a fertilizer and soil conditioner. It has been observed that sludge application reduces the surface runoff, and gives some protection against soil erosion.

The composition of sewage sludge varies greatly, depending on the source of sewage (domestic, industrial, etc.). The mean concentration level and range (percent content of dry matter) of the fertilizer content are as follows: N – 2.2 (1.5 – 3.5), P – 1.7 (1.0 – 2.5), K – 0.15 (0.05 – 0.3). The undesirable consequences of sludge application arise from the heavy metal content of the sludge, which may be taken up by the plants.

Table 3.4 compares the metal concentrations in dry sewage sludge and soil (in mg kg^{-1}) (*source:* Peter O'Neill, 1985, p. 207).

Table 3.4 Comparison of metal concentrations in dry sewage sludge and soil (in mg kg^{-1})

Metal	Mean conc. in sewage sludge	Mean conc. in soil	Conc. proportion, sludge/soil
Hg	2	0.25	8
Cd	20	0.4	50
Cu	250	50	5
Cr	500	50	10
Pb	700	25	28
Zn	3000	100	30
Fe	16,000	30,000	0.5

The metals present in sewage sludge may be divided into three categories on the basis of their availability to plants: (i) unavailable forms, such as insoluble compounds (oxides or sulfides); (ii) potentially available forms, such as insoluble complexes, metals linked to ligands, or forms attached to clays and organic matter; and (iii) mobile and available forms, such as hydrated ions or soluble complexes (Fig. 3.8). A soil which has high pH (i.e.,alkaline) and high cation exchange capacity, will be able to immobilize the metals added to it via the sewage sludge. A soil which is presently alkaline may not always remain that way. Thus, if at a later stage the pH drops (i.e., becomes acidic), the metals may get released.

Elements such as iron, zinc and copper are essential elements needed by man. The same elements may become toxic at high concentrations, however. Cadmium is a highly toxic element. Because of its geochemical affinity with zinc, it enters the plants along with zinc.

In a study made in Norway, sludge application at rates of 60–120 t (dry matter) per ha increased the Ni and Zn contents of plants, but had no significant effect on the concentration levels of Cd and Pb. Excessive

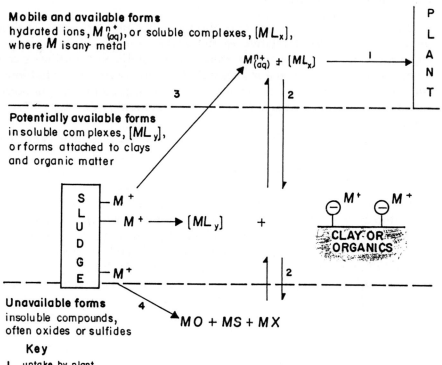

Mobile and available forms
hydrated ions, $M^{n+}_{(aq)}$, or soluble complexes, $[ML_x]$,
where M is any metal

Potentially available forms
insoluble complexes, $[ML_y]$,
or forms attached to clays
and organic matter

Unavailable forms
insoluble compounds,
often oxides or sulfides

Key

1 uptake by plant

2 changes in form due to action of
microorganisms or changes in pH or Eh

3 soluble forms pass directly into solution

4 formation of insoluble compounds

Fig. 3.8 Availablility of metals from the sewage sludge
(*source:* Peter O'Neill, 1985, p. 207) © George Allen & Unwin.

application of sludge could increase the content of heavy metals in the crops to levels which render them unfit for human consumption.

Liming (1.5 – 6 t of $CaCO_3$ ha^{-1}) reduces the heavy metal uptake by plants *in the short run*. The heavy metals will continue to persist in the top soil, however.

It has been reported that sludge-amended soils almost invariably contain higher levels of Cd than garden soils. Potential increase in the cadmium content of plants due to the application of sewage sludge constitutes the most important health hazard, which can be mitigated by increasing the available content of zinc in the soil. Sludge should not be applied to soils which are used for growing vegetables.

3.9   NPK FERTILIZERS

Application of NPK (Nitrogen, Phosphorus, Potassium) fertilizers is being increasingly preferred relative to the application of calcium nitrate, superphosphate and potassium fertilizers separately. NPK fertilizers can be custom-made to suit the nutritional status in specific soil situations, such as the deficiency of secondary nutrients (e.g. sulfur, magnesium, calcium), or micronutrients (e.g. boron and selenium) (vide Låg, 1987, for details).

(i) *Sulphur:* When a reduction in the use of superphosphate occurred globally, there was a concomitant reduction in the application of sulfur. This situation has been corrected by the use of S-bearing NPK fertilizers. The recommended rate is 16 kg of S per ha.

(ii) *Magnesium:* In the fertilizers, there exists an antipathetic relationship between K^+ and Mg^{2+}, and an affinity between K^+ and NO_3. Consequently, the enhanced uptake of potassium results in the reduced uptake of magnesium. This situation can be corrected by the addition of Mg to the fertilizers, such that agricultural areas receive about 7 kg of Mg per ha. As a consequence of the practice of addition of Mg to the NPK fertilizers, the K : Mg ratio of the fertilizers decreased from 56 to 10, and the N: Mg ratio from 60 to 16.

(iii) *Calcium:* Continuous use of NPK fertilizers leads to reduction in the Ca content of soils. Liming soils has become a common practice in several countries, which should take care of the calcium deficiency.

(iv) *Micronutrients:* NPK fertilizers are sometimes amended by the addition of boron, such that an application of about 100 g of B per ha results. In the case of deficiency of micronutrients, such as iodine, zinc and selenium, opinion differs as to the possible ways of correcting the deficiencies. In the case of iodine, it is far more effective and cheaper to add the element to a food item (in this case, as iodized table salt) than to add iodine compounds to the soil, thereby promoting the uptake of iodine by crop plants.

However, if the element concerned happens to be more valuable in the form of an organic complex, the addition of that element to the soil becomes a viable proposition. For instance, the transformation of selenium into the amino acids, methionine and cystine, promotes the ready absorption of selenium by man, and is therefore to be preferred. In Finland, sodium selenate is added to all multimineral fertilizers to offset the extremely low content of bioavailable selenium in the soils and the consequent low selenium content of grains and fodder (deficiency of selenium causes muscular dystrophy in animals, and Keshan disease, a cardiomyopathic condition, in humans). Where the fertilizers are mainly used for cereal production, the recommended supplement is 16 g of Se per ton of fertilizer (about 10 g Se ha^{-1}). In the case of fertilizers for hay and green-feed production, the recommended supplement is 6 g t^{-1}.

3.10 FERTILIZERS AND GRAIN QUALITY

Food grains vary in composition: protein (9–14%), fat (1.5 –5%), fiber (2.5–10%), and nitrogen-free extracts, mainly starch (60 –75%). Potassium and phosphorus fertilizers have only a slight effect on the chemical composition of the grains. On the other hand, nitrogen fertilizers strongly influence the protein composition. Grain protein is of two types: *active* protein which has a high content of essential amino acids and is therefore of high nutritional value, and *reserve* protein which figures in technological utilization. When higher yields are achieved due to the use of nitrogenous fertilizers, the nutritionally valuable active protein tends to be adversely affected.

Studies made in Sweden show that the grain yield increased by 130 kg (from 410 to 540 kg) when top is fertilized with 30 kg of calcium nitrate. When fertilization was increased to 60 kg calcium nitrate, production increased by only 40 kg (to 580 kg). There was hardly any increase in yield when the fertilizer application was increased beyond 60 kg. The protein content of grain increased from 8.7% (for non-N fertilizer) to 9.5 % when 30 kg of calcium nitrate was applied, and to 11% when 60 kg of calcium nitrate was applied. The nitrogenous fertilizer appears to be most effective when the last top dressing is given at the time of shooting/flowering.

3.11 FERTILIZERS AND VEGETABLES

Vegetables are a group of plants whose root-parts, stems, leaves, flowers and fruits are consumed as food in the raw or processed form. Most of the vegetables are annuals. The seeds are usually rich in proteins, fatty substances and minerals. Carbohydrates are the dominant constituents of the vegetative parts. The green parts of the plants and also root vegetables are generally rich in vitamin C and carotene.

The seed vegetables (comprising leguminous crops, such as peas and beans), and fruit vegetables require a relatively long cultivation period. Consequently, they are more vulnerable to soil nutrient imbalances. On the other hand, the leafy vegetables, despite their low nutritive value, possess several advantages: their harvesting period is flexible, a useful product can be obtained after a short period of growth, they can be dried, and their total harvest per unit of land can be very high. For instance, spinach can yield a very high amount of protein per unit area.

Since N is an integral part of chlorophyll, protein compounds, and elements of the vitamin-B group, deficiency of N in the soil affects the flavor and storage life of the vegetables, and also makes them susceptible to such diseases as rust, mildew, greymold, etc. On the other hand, excess N leads to several undesirable consequences, such as reduction in the content of useful amino acids (e.g. methionine), increase in the content of the toxic agent nitrite, and decrease in the percentage of dry matter and the biological value of the protein.

A reasonable N fertilization of spinach is $80-100$ kg ha^{-1}, but some people use as much as 300 kg ha^{-1} to increase yields; this is undesirable because of the deleterious consequences of high N in the produce (Fig. 3.9). Uncritical use of nitrogenous fertilizers has contaminated freshwater resources. Infant deaths have been reported from mathemoglobinemia, a blood disorder which arises from nitrite toxicity.

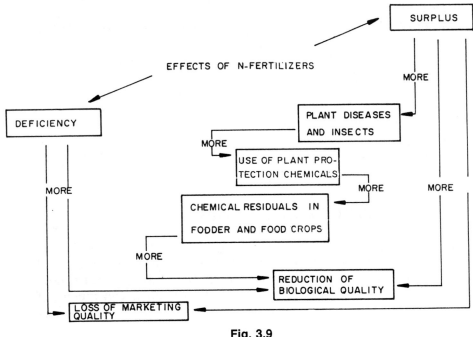

Fig. 3.9
(*source:* Schupan, 1972)

Potassium fertilization improves sugar content, taste and shelf life. Phosphorus fertilization enhances yields, but has a negligible effect on quality. The common deficiencies in micronutrients and hence the need for specific fertilization of such nutrients, varies greatly among the vegetables: boron deficiency in broccoli, cauliflower, table beet, and celeriac; copper deficiency in carrot, onion, lettuce, and spinach; zinc deficiency in onion, beans and sweet corn; molybdenum deficiency in cauliflower, etc. Deficiency or excess of any nutrient leads to impairment of the biological quality and usefulness as food and feed.

Phosphorus affects quality in particular in P-deficient areas such as Norway.

3.12 FERTILIZATION OF PASTURES AND ANIMAL HEALTH

Excessive potassium fertilization of pasture has led to magnesium deficiency and the development of hypomagnesima in dairy cows. The main site of magnesium absorption in cattle is the forestomachs. A high content of K in relation to Mg + Ca and a low Na/K ratio in the forestomachs prevents absorption of Mg in the body of the animal. To offset Mg deficiency, fertilizers are mixed with 0.2–0.3% Mg. In the case of districts with severe problem of hypomagnesima, it has been recommended that cattle be fed with mineral supplement with 13% Mg.

When forage crops are heavily N-fertilized, nitrate poisoning in ruminants occurs. The uptake of nitrate by plants is directly proportional to the nitrate content of the soil. Nitrate uptake increases under conditions of a good water supply, low pH, and low levels of molybdenum content. Nitrate itself is not very toxic, although ingestion of excess nitrate may cause gastroenteritis. When ruminants consume nitrate-bearing forage, nitrate gets biochemically altered in rumen to toxic nitrite. When the nitrite is absorbed in the bloodstream, it reacts with hemoglobin and causes the formation of methemoglobin. The body's oxygen transport system fails and the animal dies. The danger is greatest 2–3 days after the animal consumes nitrate-rich fodder. Risk of poisoning is almost certain when the nitrate content in the green fodder is more than 3% NO_3 on a dry matter basis.

Soil Degradation and Its Mitigation

And I brought you into a plentiful country, to eat the fruit thereof and the goodness thereof, but when ye entered, ye defiled my land, and made mine heritage an abomination
—Jeremiah, Old Testament.

4.1 INTRODUCTION

Soils may get degraded due to natural causes. It is not possible to control the natural causes, but it is indeed possible to mitigate their adverse consequences by being prepared for them. While anthropogenic soil degradation has taken place both in the developing and the industrialized countries, it has reached horrendous proportions in the countries of sub-Saharan Africa. Any effort to stop the degradation and roll it back needs an understanding of why and how the degradation took place in the first instance.

The German Government's Scientific Advisory Council on Global Climate Change identified a series of ailments afflicting the "skin of the earth", namely, the soil. As is the practice in the case of diseases, the ailments are named after the regions where they were first discovered by the scientists. To remedy these plagues, the Council recommended a number of "therapeutic" measures and urged that soil protection be given the same importance as accorded to climate protection.

Alps syndrome: Degradation of landscape caused by tourism.

Aral Sea syndrome: Shrinking of the surface area of water bodies due to mismanagement of water resources, and exposure of the soil to wind erosion.

Butterfield syndrome: Emissions due to automobiles and industry, resulting in acid rain and the acidification of the soil.

Dust Bowl syndrome: Soil degradation caused by industrial agriculture (in 1934, a storm in the wheat belts of central United States carried 350 million tons of dust to the East Coast).

Huang Ho (Yellow River) syndrome: Improper cultivation practices, such as cultivation on steep slopes and near the edges of stream courses (more than 1.6 billion tons of soil have been washed into the Yellow River or carried away by wind).

Katanga syndrome: Soil erosion due to mining.

Los Angeles syndrome: Extensive paving of the soil in large cities.

Sahel syndrome: Overgrazing and overexploitation of arid and semiarid land (because of this, more than 1.5 million hectares of arable land have been lost in sub-Saharan Africa in the past 30 years).

Sarawak syndrome: Soil degradation due to denudation of primary forests.

Scorched Earth syndrome: Destruction of soil due to war and weapons storage (source: *Eos*, v.76, no. 8, p.74, Feb.21, 1995).

Almost any kind of soil degradation has a geochemical component, though the criticality of such components may be highly variable.

4.2 SOIL EROSION BY WATER

Soil erosion and conservation have been comprehensively dealt with by Morgan (1986), Rattan Lal (1990) and Flanagan et al. (1991). Soil erosion leads to a decrease in a soil productivity through the physical loss of a fertile top soil, reduction in rooting depth, removal of plant nutrients, and loss of water. History is replete with examples of collapse of whole civilizations because adequate attention was not paid to the conservation of soil. It takes thousands of years for 5 cm of fertile soil layer to develop but one cm of top soil can be washed away in just one downpour. It has been estimated that annually 2 million ha of arable land is rendered unproductive due to soil erosion and degradation. Other things being equal, the rate of erosion or quantity of soil eroded may not necessarily be greater in the tropics than in temperate regions, but the erosion in the tropics does disproportionately greater harm because of the low fertility and poor quality of the exposed subsoil when the top arable layer is removed by erosion.

Soils get eroded when the impact of raindrops detaches particles from soil clods and moves them by splashing *(raindrop or rainsplash erosion)*, followed by runoff when the material thus loosened is transported by turbulent water *(runoff erosion)*. The flow may take place as unconcentrated flow in sheets *(sheet erosion)* or as concentrated flow in rills and gullies *(gully erosion)*. The extent of soil erosion is determined by the interaction between the erosivity factors and erodibility factors.

Erosivity depends on *rainfall* (drop size, velocity, distribution, angle and direction, intensity, frequency, and duration of the rain) and *runoff* (supply rate, flow depth, velocity, frequency, magnitude, duration and sediment content). *Erodibility* depends on *soil properties* (particle size, clod-forming properties, cohesiveness, aggregates, and infiltration capacity), *vegetation* (ground cover, vegetation type, degree of protection afforded by different kinds of vegetation), *topography* (slope inclination and length, surface roughness, flow convergence and divergence), and *land-use* practices (e.g. contour ploughing, gully stabilization, rotations, cover cropping, terracing, mulching, organic content). In general, erosion will be reduced if the erosivity factor is reduced, or the erodibility factor is increased (with the exception that with increase in slope inclination and slope length, the erodibility factor reduces) (Cooke and Doornkamp, 1990, pp. 80-81).

In monsoon climates, the precipitation may be confined to 2 – 3 months in a year and sometimes about half of the annual rainfall may occur in a matter of 48 h. In this situation, the raindrops tend to be big and fall with high velocities, thus causing heavy runoff carrying with it the exposed soil. There is little doubt that agriculture is the principal contributor to erosion.

Erosion is a natural process that has been going on throughout geological time. The fertile fluvisols are indeed the beneficial consequences of erosion. A steady state had been reached in the soil-water system by the time of the advent of Industrial Man. The agricultural and industrial activities of man have accelerated the rate of mechanical and chemical erosion. In the case of the tropical developing countries, deforestation, overgrazing and improper agricultural practices tend to expose the bare soil to water action and thus cause severe erosion. Erosion is caused by the following agricultural practices: farming on long slopes without terraces or runoff diversions, farming in fall-line direction on steep slopes, bare soil left after sowing of crops, bare soil left after harvest, intensive cultivation close to streams, etc. (Sytchev, 1988, p. 30).

In the case of India, of the total land area of 329 million hectares (Mha), a huge area of 129 M ha has been degraded to some extent (degraded forest land: 35 Mha; water eroded land: 74 Mha; wind-eroded land: 13 Mha; saline and alkaline land: 7 Mha.). It has been estimated that about 7 tons of soil is lost forever for every one ton of cereals produced.

The Universal Soil Loss Equation (USLE) is widely used to predict the general soil loss by rainsplash and runoff (Kirkby and Morgan, 1980):

$$A = R \times K \times LS \times C \times P,$$

where   A = average annual soil loss, R = rainfall erosivity factors ($\Sigma \, EI_{30}/100$),

K = soil erodibility factor; LS = slope length-steepness factor;

C = cropping and management factor; P = conservation practice factor.

SLEMSA (Soil Loss Estimator for Southern Africa) is a refined form of USLE and addresses the soil loss problems peculiar to Southern Africa (Stocking, 1987, Report no. 9, SADC, Lesotho).

The SLEMSA framework consists of the following:

Physical systems: Crop, Climate, Soil, Topography
Control variables: Energy interception (I), Rainfall energy (E), Soil erodibility (F), Slope Steepness (S) and Slope length (L),
Submodels: Crop ratio (C), Soil loss from bare soil (K), and Topographic ratio (X).

The SLEMSA model yields the Erosion Hazard Units (EHU) on a scale of approximately 0 – 1,000. Though the EHU figure does give a good idea of the degree of soil loss, it is not the same as t ha^{-1}. Stocking (1987) gives a number of nomograms and graphs to simplify the calculation of Erosion Hazard Units.

The most important economic consequence of soil erosion is its adverse impact on soil productivity. EPIC (Erosion-Productivity Impact Calculator) is a computer simulation model that can be used to predict the consequences of soil erosion on productivity. The model takes into account parameters of hydrology (surface runoff, percolation, subsurface flow, drainage, evapotranspiration, irrigation, etc.), weather (precipitation, temperature, radiation, wind), erosion by water and wind, nutrients (e. g. nitrogen, phosphorus, potassium), soil temperature, tillage, etc. (Williams et al., 1990).

Golubev (1980, IIASA, Austria, Rept. wp-80-129) gave estimates of the water erosion of soil in different situations (Table 4.1).

Table 4.1 Water erosion of soil in different environments (t km^{-2} y^{-1})

Zone	Province	Lowlands			Mountains		
		Past	Present	Future	Past	Present	Future
Trop.	Humid	100	810	2540	190	1840	2950
	Dry	150	330	640	280	500	780
	Arid	200	210	340	360	360	360
	Average	140	520	1430	250	1150	1810
Sub-t	Humid	80	2220	3730	210	2280	2280
	Dry	200	720	1040	390	1120	1120
	Arid	100	160	190	190	310	560
	Average	130	750	1170	280	1380	1440
Sub-b	Humid	80	2930	3120	250	6440	6440
	Dry	220	1020	1140	790	1220	1220
	Arid	60	90	110	300	300	300
	Average	140	1200	1310	390	2500	2500
Boreal	Forest	50	560	860	160	160	540
	Permafr.	30	30	100	90	90	90
	Average	40	420	660	120	120	300
World	Average	120	620	1180	250	1220	1430

Trop. = Tropical ; Sub-t = Subtropical ; Sub-b = Sub-boreal

There exist a relationship between sediment yield vis-à-vis the Mean Annual Runoff (mm) and Annual Precipitation (mm). At precipitation levels below 300 mm, erosion increases as precipitation increases. Generally, with increased precipitation, the vegetal cover will flourish better, affording better protection to the soil. Consequently, at precipitation levels of more than 300 mm, the protection afforded by the vegetation counteracts the erosive effect of greater rainfall and the soil loss decreases with increased precipitation. However, there is some evidence to suggest that at higher precipitation levels, the soil loss may increase again with precipitation, and the runoff may therefore increase (Morgan, 1986, p. 4; Sytchev, 1988, p. 31).

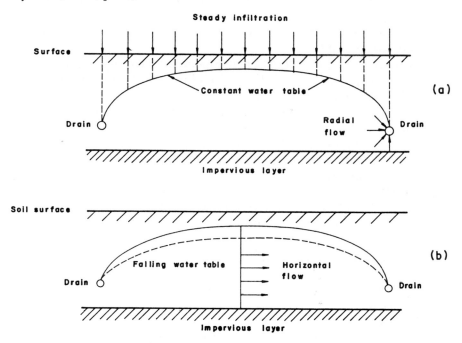

Fig. 4.1   Groundwater drainage under steady and nonsteady flow conditions (*source:* Hillel, 1987, p. 85 © World Bank).

4.3   SOIL EROSION BY WIND

Analogous to the Universal Soil Loss Equation (USLE), a wind-erosion equation was developed (Woodruff and Siddoway, 1965, as annotated by Cooke and Doornkamp, 1990, p. 248):

$$E = f\,(I', K', C', L', V'),$$

where   E = erosion (per hectare, per annum),

I' = comprises the soil erodibility index, I — the potential loss from a wide, unsheltered, isolated field with a bare, smooth, noncrusted surface; and the knoll erodibility index (I_s), or erodibility of windward slopes expressed as percentage slope,

K = Soil ridge roughness factor, which is a measure of natural or artificial roughness other than that caused by clods and vegetation,

C' = Local wind climatic factor,

L' = field length or equivalent along the prevailing wind-erosion direction, based on the total distance across the field measured along the prevailing wind-erosion direction (D_f) and the sheltered distance in the same direction (D_b),

V' = Equivalent quantity of vegetal cover based on R, the surface residue, S, the total cross-sectional area of vegetal cover and K_o the measure of vegetal roughness.

Because of the complexities involved, the usual practice is to solve the equation graphically or by use of computers.

Wind erosion selectively removes small-diameter, low-density particles from the soil exposed on the ground surface, either because of ploughing or the removal of vegetation. The erodibility of soil due to wind depends on the roughness of the ground surface and the extent of plant cover. An increment in wind velocity will increase the range of particle size and density of the material carried away. Individual soil particles may move by a bouncing motion called saltation, and fine particles are carried in suspension in the form of a dust cloud.

4.4   SOIL, DROUGHT, AND DESERTIFICATION

There is a well-established link between precipitation and ecosystems, to wit:

Ecosystem	Precipitation (mm/y)
Desert	< 100
Arid lands	< 250
Semiarid lands (rangelands)	> 250

Semiarid tropical lands are characterized by long dry seasons, low and unpredictable rainfall and poor soils.

Falkenmark et al. (1990) emphasize that droughts and famines should not be regarded as disaster *events* but *processes*. Drought occurs when the available moisture in the root zone falls below 30% of the water-holding capacity of the soil. They distinguish four different kinds of water scarcity:

Natural: (i) aridity, reflected in the short length of the growing season, and (ii) intermittent droughts, reflected in recurrent drought years with risk of crop failure;

Man-made: (iii) landscape desiccation caused by soil degradation, resulting in water not reaching the plant roots, and (iv) water stress, whereby the demand for water by the growing population exceeds the regenerative capacity of the system.

Desertification is land degradation in arid, semiarid and dry subhumid areas. It may be caused by various factors, notably climatic variations and human activities. By leading to a reduction in growth of important plant species, droughts can initiate or accelerate desertification. Grazing can accentuate the process by stripping the land of its cover. However, plant growth depends not only on precipitation, but also on other parameters such as temperature, soil moisture capacity, and species. If the environment becomes drier and the soil further degraded through erosion and compaction, irreversible desertification could set in.

Previously, desertification used to be defined only in terms of paucity of rainfall. Presently, however, desertification is defined as the environmental degradation arising from human and climatic causes *under any rainfall regime*. It has been estimated that 60,000 km^2 y^{-1} of land is lost due to desertification.

Overgrazing and removal of the vegetal cover are the principal causes of desertification. Wind is the principal agent responsible for the erosion of exposed soil in arid lands. As wind preferentially removes the clays with which the essential nutrient elements are associated, wind erosion drastically reduces soil fertility.

During drought years the environment is stressed to such an extent that even the limited vegetation that exists is used up either as fuel or fodder. This has happened more in sub-Saharan Africa than in any other part of the world, so much so that about one-fourth of the land in sub-Saharan Africa has already been desertified almost irrevocably. Some types of grasses, shrubs and acacia trees could be grown even in such harsh conditions, however.

4.5 LOSS OF FARMLAND DUE TO URBANIZATION

History shows that most of the large cities in the world grew from small towns and villages which were originally sited in rich fluvisols in river valleys or along the coast. Every year, about 8 million hectares of good agricultural land is lost due to urbanization (construction of habitations, industries, roads, etc.). It has been estimated that by the year 2000, 155 million ha of presently highly productive cropland and grassland,

440 million ha of forest land, 200 million ha of potentially productive cropland, would have been lost, besides the degradation of about 100 million ha of land.

4.6 SOIL ACIDITY AND ITS MITIGATION

A soil is said to be acidic if the pH of its aqueous solution phase is less than seven. Soils in the humid tropics tend to be acidic, because they form as a consequence of weathering which takes place under conditions of intensive leaching by fresh water (fresh water invariably contains free protons at concentration levels of more than 1 mmol m^{-3}). Acid soils are also common in the forested areas in the temperate zones.

The range of pH in soils is 3 to 9. pH values of most cultivated soils has a range of 5.5 to 7.5. Tea grows well at pH of 4.5, and the optimal pH for wheat is 6 – 7.5.

High soil acidity has a severe adverse effect on plant growth. It hinders the activities of micro- and mesoorganisms, such as bacteria and earthworms, which prefer conditions around neutrality. There is an increase in the phytotoxic elements, such as Mn and Al, in the soil solution. Formation of ammonia and nitrate is inhibited. Other deleterious consequences are decrease in the availability of phosphorus, deficiency of Mo, reduction in root activity and increase in the propensity for certain diseases like club root.

The speciation of various elements in an acidic solution (pH 4.7) is given in Table 4.2 (*source:* Sposito, 1989, p. 217).

Table 4.2 Speciation of an acid soil solution (pH 4.7)

Constituent	C_T (mmol m^{-3})	Percentage speciation
Ca	20	Ca^{2+} (99%), $(CaSO_4)^0$ (1%)
Mg	6	Mg^{2+} (99%), $(MgSO_4)^0$ (1%)
K	3	K^+ (100%)
Na	20	Na^+ (100%)
Al	17	AlL (71%), AlF^{2+} (11%), Al^{3+} (11%), $Al(OH)^{2+}$ (5%), $Al(OH)_2^+$ (2%)
CO_3	10	$H_2CO_3^*$ (100%)
SO_4	54	$(SO_4)^{2-}$ (99%)
Cl	24	Cl^- (100%)
F	2	AlF^{2+} (95%), F^- (5%)
SiO_2	100	$Si(OH)_4^0$ (100%)
L*	12	AlL (100%)

* Organic ligands from soil humus

Three factors play an important part in producing proton concentration in a soil solution and making it acidic: (i) *wetfall* (precipitation in the form of rain, snow, fog drizzle), (ii) *dryfall* (deposits of solid inorganic and organic particles, of natural or anthropogenic origin), and (iii) *interflow* (lateral movement of soil solution in the subsurface). Protons are lost from the system through the processes of volatilization, wind erosion, etc. Anthropogenic inputs seriously perturb the natural equilibrium through the industrial effluents (e.g. SO_x, NO_x) that produce the wetfall, and the nitrogenous fertilizers which lead to acid conditions. The specific processes that influence the pH of the soil solution are: dissociation of carbonic acid ($H_2CO_3^-$), interaction between the soil humus and aluminum hydroxy polymers, and mineral-weathering reactions.

The pH of the soil solution is determined by the partial pressure of CO_2 gas (in atmospheres) and the bicarbonate ion activity (mmol m^{-3}):

$$P_{CO_2}/(H^+)(HCO_3^-) = 10^{7.8} \ (T = 298.15 \ K) \tag{4.1}$$

The equation may be rewritten in the logarithmic form as:

$$pH = - \log(H^+) = 7.8 + \log(HCO_3^-) - \log P_{CO_2} \tag{4.2}$$

For soil air in B horizons or near plant roots, $P_{CO_2} = 10^{-2}$ atm., and pH = 4.9. Thus, soil solutions can be expected to have of a pH of about 5.0 if the dissociation of carbonic acid happens to be the controlling chemical reaction.

"The acid-neutralizing capacity (ANC) is the moles of protons per unit volume or the mass required to change the pH value of an aqueous system to the pH at which the net charge from ions that do not react with OH^- or H^+ is zero (Sposito, 1989, p. 211). ANC can be expressed in the following form:

$$ANC = [HCO_3^-] + 2\,[CO_3^{2-}] + [OH^-] - [H^+] \tag{4.3}$$

ANC increases with increasing pH. The rate of change of ANC with pH (d ANC/d pH) is called the buffer intensity, β_H. The buffer intensities of organic-rich, temperate-zone acid soils have a range of 0.1 – 1.5 mol_c kg_{om}^{-1} pH^{-1} (*om* is organic matter).

Aluminum hydroxy polymers — whether in aqueous solution, adsorbed on soil particles or in solid phases — profoundly affect the pH of the soil solution in the mineral horizons of the acid soils. The biological processes which influence the soil acidity are the uptake or release of ions by the plant roots and the microbial catalysis of redox reactions. When plants pick up more cations from the soil than anions, the rhizosphere may become more acid than the bulk soil (sometimes, by as much as two units of pH).

The proton budgets are estimated in terms of kmol ha^{-1} yr^{-1} in the field soils. Protons are produced by wetfall and dryfall, CO_2 (aq) and organic matter, and biouptake and release reactions (involving C, N and S). Protons are consumed in the mineral-weathering reactions (such as, hydrolysis, complexation and ion exchange) or may be lost due to interflow processes or deep percolation. If the production of protons is more than the consumption or loss of protons, as is most often the case in tropical areas and sometimes in Podzols in the temperate regions, the soil becomes progressively acid.

Neutralization of soil acidity

Liming is the time-honored method of mitigating soil acidity. Liming supplies calcium, improves the solubility of phosphorus, and precipitates aluminum.

The lime requirement is stated in terms of moles of Ca^{2+} charge per kilogram of soil required to decrease the total acidity to a value deemed acceptable to the agricultural use of the soil (pH around 6). It is conveniently expressed in units of mol$_c$ kg^{-1}.

The basic chemical reaction related to the use of liming of acid soils may be simplified as follows:

$$2\ Al\ X_3\ (s) + 3\ CaCO_3\ (s) + 3\ H_2O\ (l) = 3\ CaX_2\ (s) +$$
$$2\ Al\ (OH)_3\ (s) + 3\ CO_2\ (g) \qquad (4.4)$$

where $\quad$ X = one mol of negative intrinsic surface charge; s = solid; l = liquid; g = gas.

The amount of lime required to ameliorate an acid soil will depend upon the mineralogy of the soil, content of clay and organic matter, the extent of leaching with fresh water, and where relevant, proton inputs from acid deposition and acidifying fertilizers. The lime amendment is generally in the range of 1.5 – 6 t of $CaCO_3$ ha^{-1}. Adams (1984) gave a detailed account of the liming practices in agriculture.

4.7 SALT-AFFECTED SOILS AND THEIR RECLAMATION

The gravity of the soil salinization problem in the global context can be appreciated from the fact that salt-affected soils occupy an estimated 952 Mha (million hectares), which constitute about 7% of the total land area, or nearly one-third of the potential arable land in the world (Gupta and Abrol, 1990). It has been estimated that about 125,000 ha of irrigated land per year becomes unproductive due to salinization. For instance, about 7 million ha of land in India has become saline due to improper agricultural practices. The economic costs of irrigation-induced soil salinization are mind-boggling. For instance, the World Bank estimates that soil salinity is "robbing" Pakistan of USD 2.5 billion annually in lost agricultural productivity.

Gupta and Abrol (1990) gave an excellent account of the procedures for the reclamation of salt-affected soils. Farm forestry has great potential to ameliorate salinity problems. Experience in the Punjab province (India) has shown that through the scientific management of soil, water, and crop, it is indeed possible to make the salt-affected soils productive.

4.7.1 Salinity and Sodicity

Salinity is a characteristic feature of arid-zone soils. In the arid zones, evaporation is greater than precipitation. Consequently, when the soils dry, ions released into the soil solution (by mineral weathering, or by leached salts in tropical soils coming to the surface by capillary action or by the intrusion of saline groundwater) accumulate in the secondary minerals, such as clays, carbonates, sulfates and chlorides. Metal cations, such as Na, K, Ca and Mg, can be readily brought into solution, either in the form of exchangeable ions on smectite and illite or as component ions in carbonates, sulfates and chlorides. These ions combine with ligands, such as CO_3, SO_4 and Cl, to form the characteristic mineral assemblages of arid-zone soils.

Exchange reactions among the cations Na^+, Ca^{2+}, and Mg^{2+} is a characteristic feature of arid-zone soils. Such reactions cannot be expressed as straightforward chemical equations because of the extreme heterogeneity of soil as an adsorbent. So the equation given below may be deemed to be an analogue of the conventional chemical equation, and approximates the ion exchange characteristics of a soil.

$$Na_2X(s) + Ca^{2+} (aq) = CaX(s) + 2\ Na^+ (aq) \qquad (4.5)$$

where X represents 1 mol of negative exchanger charge.

In the case of saline soils, sodicity becomes an important parameter. The *Sodium Adsorption Ratio* (SAR) is given by the equation,

$$SAR = c_{Na}/\sqrt{C_{ca}} \text{ , where } C \text{ is in mol m}^{-3} \qquad (4.6)$$

Since the direct measurement of Na^+ and Ca^{2+} is not a practical proposition, SAR is replaced by the parameter, SAR "practical" or SAR_p.

$$SAR_p = Na_T/\sqrt{Ca_T} \qquad (4.7)$$

where Na_T and Ca_T represent the total concentrations. Since Ca^{2+} has a greater tendency for complex formation than Na^+, SAR_p is about 12% lower than SAR on average. Since Na-Mg exchange may also be involved, SAR_p can be conveniently measured as follows:

$$SAR_p = Na_T/(Ca_T + Mg_T)^{1/2} \qquad (4.8)$$

SAR_p is roughly equal to the parameter called Exchangeable Sodium Percentage (ESP).

The salinity of a soil is measured in terms of the electrolytic conductivity (EC_e) of its aqueous phase, obtained by extraction from a saturated paste. It is expressed in units of decisiemens per meter ($dS\ m^{-1}$) or microsiemens per cm ($\mu S\ cm^{-1}$). A soil is said to be saline if its EC_e is > 4 $dS\ m^{-1}$.

The *Marion-Babcock* equation gives the relationship between the electrolytic conductivity (k, in units of $dS\ m^{-1}$) and ionic strength (l, in units of $mol\ m^{-3}$):

$$\log l = 1.159 + 1.009 \log k \qquad (4.9)$$

Thus, EC_e of 4 $dS\ m^{-1}$ corresponds to the ionic strength of 58 $mol\ m^{-3}$. This level of salinity is only 10% of that of sea water, but even salt-tolerant crop species are affected at EC_e of 2 $dS\ m^{-1}$ (29 $mol\ m^{-3}$). Agricultural crops generally do not tolerate salinities greater than 1 $dS\ m^{-1}$ or 14 $mol\ m^{-3}$.

The alkalinity of a soil solution is defined by the equation:

$$Alk = [HCO_3^-] + 2\ [CO_3^{2-}] + [OH^-] - [H^+] \qquad (4.10)$$

The pH value of a saline soil solution whose alkalinity mainly comes from bicarbonate, is given by the equation:

$$pH = 7.8 + \log (HCO_3^-) - \log P\ CO_2 \qquad (4.11)$$

The chemical speciation of alkaline soil solution (at pH 7.6), representing a saturated extract from an irrigated Aridisol, is given in Table 4.3 (*source:* compilation by Sposito, 1989, p. 227):

Attention is drawn to some features of the Table: (i) soluble Ca dominates over soluble Mg, (ii) Ca, Mg, HCO_3^-, and PO_4, are characterized by complicated speciation, (iii) organic complexes of metal cations are unimportant, and (iv) Na, K, Cl, and NO_3 have very high free ion percentages.

4.7.2 Why Drainage ?

In the context of agriculture, drainage refers to artificial removal of excess water from cropped fields. Drainage may be of two kinds, surface drainage and groundwater drainage. Excess water accumulating on the surface is drained by shaping the land to facilitate overland flow (surface drainage). Excess water within the soil or subsoil is drained by lowering the water table or preventing its rise (groundwater drainage).

While soil saturation for short periods is not harmful *per se*, prolonged saturation (waterlogging) can adversely affect plant growth. Plants need to respire constantly. Excess water in the soil blocks the soil pores, impedes movement of oxygen from the atmosphere, and thereby severely

Table 4.3 Speciation of an alkaline soil solution (pH 7.6)

Constituent	C_T (mmol m^{-3})	Percentage speciation
Ca	5.9	Ca^{2+} (82 %), $CaSO_4^0$ (16%), $CaHCO_3^+$ (1%)
Mg	1.3	Mg^{2+} (85%), $MgSO_4^0$ (13%), $MgHCO_3^+$ (1%)
Na	1.9	Na^+ (99%), $NaSO_4^-$ (1%)
K	1.0	K^+ (98%), KSO_4^- (1%)
HCO$_3$	3.0	HCO_3^- (91%), $H_2CO_3^0$ (4%), $CaHCO_3^+$ (3%), $MgHCO_3^+$ (1%)
SO$_4$	4.4	SO_4^{2-} (74%), $CaSO_4^0$ (21%), $MgSO_4^0$ (4 %)
Cl	5.0	Cl^- (99%), $CaCl^+$ (1%)
NO$_3$	0.28	NO_3^- (99%), $CaNO_3^+$ (1%)
PO$_4$	0.065	HPO_4^{2-} (36%), $CaHPO_4^0$ (35%), $MgHPO_4^0$ (13%), $H_2PO_4^-$ (8%), $CaPO_4^-$ (4%), $MgPO_4^-$ (1%)
B	0.038	$H_3BO_3^0$ (96%), $B(OH)_4^-$ (3%)
L *	0.022	L^- (95%), CaL^+ (4%), MgL^+ (1%)

* Organic ligands from soil humus.

limits the respiration activity of plants. When the soil thereby becomes anaerobic, reducing conditions may prevail, leading to the development of toxic concentrations of ferrous and manganous ions as well as sulfides and the formation of methane. Nitrification is prevented and denitrification may consequently occur. Besides, wet soils allow the growth of harmful root pathogens, such as fungi (Hillel, 1987, p. 81).

Decrease in solubility of oxygen in water and increase in respiration rate of both plants and microorganisms result in temperature rise. Thus, waterlogging has more severe consequences in warm climates than in cold climates.

The application of too much water and lack of drainage would raise the water table, and the leached salts would come to the surface by capillary action, thus rendering the soil saline. In medium-textured soils, the upward flow of salt-bearing water during the dry season may be one to two meters.

Thus, it follows that all *irrigated lands must have drainage*. Irrigation without drainage can have disastrous consequences. Ancient civilizations based on irrigated agriculture in river valleys (e.g. Mesopotamia) collapsed, as they did not provide for drainage. The land became infertile due to steady rise of the watertable, waterlogging, and salinization. The salinization problem persists in many river valleys in the world, such as the Indus (India and Pakistan), Nile (Egypt), Murray (Australia), Colorado (USA), etc. Groundwater drainage is undoubtedly expensive, but it should always be planned for right from the beginning. Reclamation of saline land is prohibitively expensive, and in some cases, even impossible.

4.7.3   Methods of Subsoil Drainage

Excess water from the soil or subsoil is drained through ditches, perforated pipes, or machine-formed "mole" channels, by gravity flow or pumping. The water thus drained is let into a stream or a lake or an evaporation pond or the sea. As the drainage water invariably contains undesirable concentrations of harmful salts, fertilizer nutrients or pesticide residues, it should never be let into a river whose water will be used for drinking purposes by people on the downstream side. The drainage water is often recycled or reused for agricultural, industrial or recreational purposes. For instance, the drainage water from the fertile Imperial Valley of California, USA, is let into a lake (Salton Sea) where it is used for recreational purposes (e.g. boating and water-skiing).

The rate of flow from the soil to the drains depends upon the following factors: (i) permeability of the soil, (ii) depth of watertable, (iii) depth of drain, (iv) horizontal spacing between drains, (v) configuration of the drains (open ditches or tubes), their diameter and slope, (vi) nature of the drain-enveloping material (such as gravel) used to increase the seepage surface, and (vii) rate of recharge of groundwater (i.e., excess of infiltration over evapotranspiration) (Hillel, 1987, p. 83).

As the field conditions tend to be highly complex and variable, recourse has to be taken to the use of empirical equations to estimate the desirable depths and spacing of drain pipes under particular soil and groundwater conditions.

The classical equation of Hooghoudt can be used for the purpose:

$$H = QX (S - X)/2 KD \qquad (4.12)$$

where H = height above the drain, X = horizontal distance to the nearest drain, Q = percolation flux (the excess of infiltration over evapotranspiration), S = spacing between adjacent drains, K = hydraulic conductivity of the saturated soil, D = height of the drains above the impervious layer that is presumed to exist at some depth in the subsoil.

It can be seen from the equation that the height of rise of watertable between the drains is directly related to the recharging flux and to the square of the distance between the drains, and inversely related to the hydraulic conductivity of the soil (Hillel, 1987, p. 84).

Typical depths and spacing of tile drains for different soil types are given in Table 4.4 (*source:* Hillel, 1987, p. 84).

In Holland, a country which has developed drainage into a fine art, drainage seeks to remove about 7 mm d^{-1} and prevent the watertable rise above 50 cm from the soil surface. In arid regions, the watertable

Table 4.4 Typical depths and spacing of tile drains for different soil types

Soil type	Hydraulic conductivity (cm d^{-1})	Spacing of drains (m)	Depth of drains (m)
Clay	0.15	10 - 20	1 - 1.5
Clay loam	0.15 - 0.5	15 - 25	1 - 1.5
Loam	0.5 - 2.0	20 - 35	1 - 1.5
Sandy loam	6.5 - 12.5	30 - 70	1 - 2
Peat	12.5 - 25	30 - 100	1 - 2

must be kept considerably deeper because of greater evaporation and consequent faster rate of increase in salinity.

Figure 4.1 (*source:* Hillel, 1987, p.85) shows the groundwater drainage under steady and nonsteady flow conditions (see p. 84). The drainage layout for a 25 hectare farm unit is depicted in Fig. 4.2.

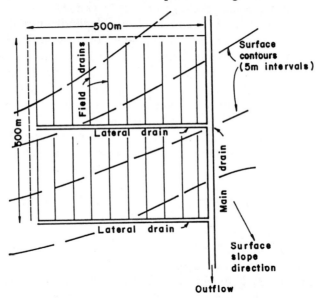

Fig. 4.2 Drainage layout for a 25 ha farm unit
(*source:* Hillel, 1987, p. 85 © World Bank).

4.7.4 Irrigation Water Quality and Salinization

Chemically, irrigation waters are electrolyte solutions. According to Shainberg and Oster (1978), the quality of irrigation water is determined on the basis of the following criteria: (i) total salinity (total concentration of all salts in the water), (ii) sodicity (concentration of sodium relative to other cations), (iii) anion composition, particularly the content of carbonate and bicarbonate ions, and (iv) concentration of toxic elements, chiefly boron.

Irrigation waters drawn from surface or groundwater sources typically contain 200 to 2,000 ppm of salts (or 200 – 2,000 g of Total Dissolved Solids per m^3). Thus, the application in a single season of 1,000 mm of medium quality water (containing, say, 500 ppm of salts), would add 5 tons of salt to a hectare of land. If salt is allowed to accumulate at this rate, the land will become saline and infertile in a matter of a few seasons. *Hence, the imperative need for leaching and drainage to remove the salt, especially if the irrigation water is even slightly brackish.*

As the demand for irrigation water of good quality is in excess of supply of such water, it is inevitable that water of inferior quality (in terms of dissolved salts and contamination) has to be used. This involves either preprocessing of water before application, or growing crops which can tolerate inferior quality water. Shuval et al. (1986) gave an account of using waste water for irrigation. Use of brackish water for irrigation was found to be entirely feasible under certain conditions (growing tolerant crops and high-frequency irrigation) (Hillel, 1987, p.89). There are a number of saline-tolerant plants which can be grown in the saline coastal wastelands and yield valuable nonedible oils and products usable in industries. Jojoba (*Simmondsia chinensis*) provides seed oil (50 – 55%) similar to sperm whale oil. *Salvadora persica* yields 35 – 40% seed oil rich in lauric acid which can replace coconut oil in soap and detergent industries. *Atriplex nummularia* is a potential biological desalinator. *Juncus rigidus* is a source of paper pulp. Some commercially important plant species are being biogenetically altered to possess high salinity tolerance (vide brochure of CSMCRI, Bhavnagar, India).

The chemical composition of the irrigation waters would depend upon the source and postwithdrawl treatment. Before using the water for irrigation, it is necessary to project how the chemistry of irrigation waters will affect the kind of compounds and weathering processes that exist in the soil. The impact of irrigation water quality on irrigated agriculture is studied in terms of salinity hazard, sodicity hazard, and toxicity hazard.

Soils may have primary salinization arising from the content of residual salts, or secondary salinization may be caused by natural or human causes. Soils may be rendered saline either due to insufficient downward movement, application of poor quality irrigation water, or the groundwater being too close to the surface. Kovda found that even if good quality irrigation water is used, salinization could still take place if the watertable were to rise above the critical depth for a given soil.

The critical depth (L, in cm) can be computed from the following formula of Kovda (1980, p.205):

$$L = (17 \times 8 \ t) + 15, \hspace{3cm} (4.13)$$

where t is the average annual temperature in °C.

Crops can be broadly divided into categories, depending on their tolerance of salinity: Sensitive crops require the application of irrigation water with electrolytic conductivity (EC_w) of less than 0.7 dS m^{-1} (equivalent to ionic strength <10 mol m^{-3}). Relatively tolerant crops can withstand up to 3 dS m^{-1} (44 mol m^{-3}). Water with EC_w > 3 dS m^{-1} (> 44 mol m^{-3}) should not ordinarily be used for irrigation, except with appropriate amendment.

It should be borne in mind that soil salinity is defined in terms of EC_e and not in terms of EC_w of applied water. The relationship between EC_e and EC_w in the root zones is complex because it depends on many factors. The usual practice therefore is to use empirical "rule of thumb".

The steady-state value of EC_e resulting from the application of water with conductivity EC_w can be estimated from a knowledge of the leaching fraction (LF).

$$LF = \frac{\text{Volume of water leached below root zone}}{\text{Volume of water applied}} \qquad (4.14)$$

Typically, LF is in the range of 0.15 – 0.20. This would mean that 15 – 20% of the water applied leaches below the root zone, whereas evapotranspiration accounts for 80 – 85% of the applied water.

EC_e in the root zone can be computed from EC_w of applied water, from the following equation:

$$EC_e = X \, (LF). \, EC_W \qquad (4.15)$$

where X (LF) is an empirically estimated parameter based on experience with irrigated, cropped soils. Ayers and Wescot (1985) gave X(LF) factors for different values of LF. For instance, if the EC_w is 1.5 dS m^{-1}, and LF is 0.20 (with the corresponding X (LF) factor of 1.3), then EC_e would be 1.95 dS m^{-1}. For LF values greater than 0.3, the X (LF) value will be less than 1.0, and EC_e becomes less than EC_w. Application of water with EC_w > 2 dS m^{-1}, under conditions of LF < 0.1, will render the soil saline.

Table 4.5 and Fig. 4.3 give the irrigation water quality criteria of the U.S. Department of Agriculture (USDA Handbook 60).

4.7.5   Tolerance of Crops to Alkaline and Saline Soil Environments

The crops to be grown during the reclamation of alkaline soils have to be carefully chosen to obtain acceptable yields. Sodicity tolerance ratings for different crops are fixed on the basis of crop hazard as manifested by 50% reduction in relative yields (Gupta and Abrol, 1990, p. 263).

On the basis of extensive field studies, the Central Soil Salinity Research Institute, Karnal, India, gave the relative crop tolerance to alkalinity/sodicity of soils in terms of ESP (Exchangeable Sodium Percentage) range:

Table 4.5 Salinity and sodicity of water usable for irrigation

Salinity/sodicity class	Description and use
C1	Low salinity water which can be used for irrigation with most crops on most soils.
C2	Medium salinity water which can be used for plants with moderate salt tolerance.
C3	High salinity water which cannot be used on soil with restricted drainage.
C4	Very high salinity water which is unsuitable for irrigation, except where the soils are highly permeable and high salt-tolerant crops are to be grown.
S1	Low sodium water can be used for irrigation of most soils. Sodium-sensitive crops may still accumulate injurious levels of sodium.
S2	Medium sodium water, presents hazard for fine-textured soils; may be used on coarse-textured or organic soils with good permeability.
S3	High sodium water unsuitable, except in the case of gypsiferous soils.
S4	Very high sodium water generally unsatisfactory, except where gypsum amendment is made.

ESP range	Crops
10-15	Safflower, mash, peas, lentil pigeon-pea, urd bean
16-20	Bengal gram, soybean
20-25	Peanut, cowpea, onion, pearl millet
25-30	Linseed, garlic, guar
30-50	Raya, wheat, sunflower
50-60	Barley, sesbania
60-70	Rice

Crops have widely varying levels of salt tolerance — glycophytes are highly sensitive to salt and tolerate only low levels of salinity, whereas halophytes can tolerate higher levels of salt concentration. Plant response to salinity depends not only on salt concentration level, but also upon a variety of factors, such as temperature, humidity, light intensity, stage of growth, moisture, soil fertility, etc. For this reason, it is not possible to prescribe universally applicable, crop-specific salt tolerance limits. Nevertheless, a broad idea of the degree of salt tolerance of crops is useful in agronomic planning.

Salt tolerance of crops is judged on the basis of two criteria: (i) threshold soil salinity (Ct), which leads to no reduction in yield. This parameter is expressed in terms of electrolytic conductivity of saturated paste extract, in dS m^{-1}, and (ii) slope (S), which corresponds to decrease in percentage yield per unit of salinity increase beyond the threshold.

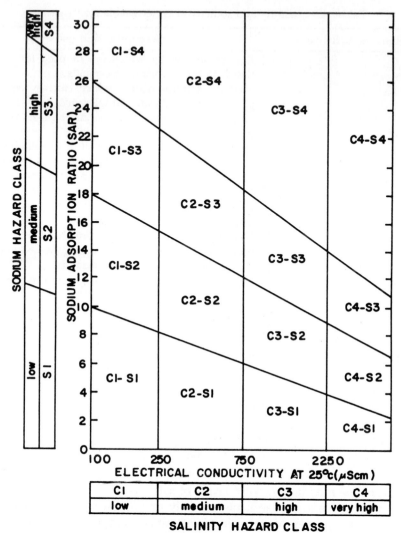

Fig. 4.3 Irrigation water quality criteria
(*source:* U.S. Department of Agriculture).

Table 4.6 gives the salt tolerance of selected crops, grasses and trees
(*source:* Maas, 1988)

The decrease in relative yield (YR) of crops due to increase in salinity
is given by the following equation of Mass and Hoffman (1977):

$$YR = 100 - S\,(C - Ct) \qquad (4.16)$$

where C is the average salinity of the root zone.

Table 4.6 Salt tolerance of selected crops, grasses and trees

Crop	Salt tolerance threshold (dS m^{-1})	50% yield (dS m^{-1})
Sensitive		
Tomato	0.5	7.6
Onion	1.2	4.2
Lettuce	1.3	5.2
Cowpea	1.3	9.1
Berseem	1.5	9.5
Corn	1.7	5.9
Sugarcane	1.7	9.8
Spinach	2.0	8.5
Alfalfa	2.0	9.0
Sesbania	2.3	9.3
Rice	3.0	7.2
Peanut	3.2	5.0
Beet, red	4.0	9.6
Date palm	4.0	16.0
Tolerant		
Asparagus	4.1	29.0
Squash	4.7	9.9
Soybean	5.0	7.5
Wheat	6.0	13.0
Sorghum	6.8	15.0
Bermuda grass	6.9	14.8
Sugarbeet	7.0	15.0
Cotton	7.7	18.0
Very tolerant		
Barley	8.0	18.0
Guayule	15	19

Slope (S) can be calculated from the following expression:

$$S = 50/(EC_{50\% \text{ yield}} - EC_{\text{threshold}}) \qquad (4.17)$$

Though the field conditions may depart considerably from the steady-state condition assumed in the equations, the threshold and slope parameters are undoubtedly useful in modeling irrigation with brackish water.

4.7.6 Salt Balance and Leaching Requirement

The change in the salt content (S) in the root zone is given by the excess of the input of salt over the output.

Salt content change = [input of salt] – [output of salt]

$$S = [(Vi)\,(Ci) + (Vr)\,(Cr) + (Vc)\,(Cc)] - [(Vd)\,(Cd) - (Vp)\,(Cp)] \qquad (4.18)$$

where Vi, Vr, Vc are respectively the volumes of water entering the root zone as irrigation (i), rainfall (r), and capillary rise (c), and Vd and Vp are volumes of water leaving the root zone as drainage (d) and plant uptake (p). Ci, Cr, Cc, Cd, and Cp are corresponding average concentrations of salt in the same volumes of water. The terms Cr and Cp are negligible, as the salt concentrations in rain water are low (except near the coasts where the rain water may be affected by salt spray), and the plants hardly take up any salts. Thus, the above equation could therefore be simplified as:

$$S = (Vi)\,(Ci) - (Vd - Vc)\,(Cd) \qquad (4.19)$$

If the watertable is kept sufficiently deep, such that the capillary rise of groundwater to the root zone is effectively prevented, Cd becomes negligible and can be ignored.

The condition for non-accumulation of salt in the root zone is that S should be equal to zero.

Thus, $(Vi)\,(Ci) = (Vd)\,(Cd),$ or $Vi/Vd = Ci/Cd$ (4.20)

It may be noted that the volume of water drained, Vd, is equal to the difference between the volumes of irrigation, Vi, and of evapotranspiration, Vet. Customarily, water volumes are expressed in terms of mm of depth ($10\ m^3\ ha^{-1}$ is equivalent to a depth of one mm).

On the basis of the above considerations, Richards (1954) gave the following equation:

$$Di = [Cd/(Cd - Ci)]\,Det, \qquad (4.21)$$

where Di is the depth of irrigation, and Det is equivalent depth of water consumed by the plant, i.e., evapotranspiration.

The US Salinity Laboratory developed the concept of "leaching requirement" as a monitoring mechanism to ensure that no excessive build-up of salt takes place in the root zone of the plant. The leaching requirement has been defined as the fraction of the irrigation water that must be leached out of the bottom of the root zone in order to prevent average soil salinity from rising above some permissible limit (Richards, 1954). Thus, the leaching requirement is determined by the salt content of the irrigation water, rate of evapotranspiration, and the specific salt tolerance of the crop concerned (vide Table 4.5).

4.7.7 Reclamation of Alkaline Soils

In some countries, intensive agriculture has accentuated the degradation of soils by salinization and alkalization. It has been found that saline and alkaline soils tend to occur in distinct geographic zones. Areas of occurrence of alkaline soils are characterized by a high content of exchangeable sodium in the soil profile, and a low total salinity and significant residual alkalinity of the groundwater.

Alkaline soils are reclaimed by the process of amendment which involves removing part or most of the exchangeable sodium (which is harmful to plants), and its replacement by the more favorable calcium ions in the root zone. Agronomic practices such as rice cropping and use of low-nitrogen organic manures, facilitate the process. The chemical amendment is to be followed by leaching for the removal of soluble salts and other reaction products of the amendment.

Chemical amendments for the reclamation of alkaline soils are grouped into three categories: (i) soluble calcium salts, such as gypsum, calcium chloride and phospho-gypsum, (ii) sparingly soluble calcium salts, such as calcium carbonate, and (iii) acids or acid formers, such as, sulfuric acid, iron and aluminum sulfates, lime-sulfur, pyrites, etc. (vide Table 4.7, *source:* Gupta and Abrol, 1990). All amendments are meant to supply soluble Ca; calcium salts do this directly and acid amendments indirectly, by reacting with lime naturally present in the soil. The atmospheric oxidation of sulfur and pyrites to produce sulfuric acid is a very slow process, unless accelerated by such microbes as *Thiobacillus ferroxidans*.

Table 4.7 Chemical characteristics of various amendments

Amendment	Chem. Comp.	Gypsum (kg)*	Sulfur (kg)*
Gypsum	$CaSO_4 . 2H_2O$	1.0	5.38
Sulfur	S_8	0.19	1.00
Sulfuric acid	H_2SO_4	0.57	3.06
Calcium carbonate	$CaCO_3$	0.58	3.13
Calcium chloride	$CaCl_2 . 2H_2O$	0.85	4.59
Ferrous sulfate	$FeSO_4 . 7H_2O$	1.61	8.69
Ferric sulfate	$Fe_2(SO_4)_3 . 9H_2O$	1.09	5.85
Aluminum sulfate	$Al_2(SO_4)_3 . 18H_2O$	1.29	6.94
Pyrite (30% S)	FeS_2	0.63	1.87

* Amount equivalent of 1 kg of chemically pure substance.

Gypsum is easily the most suitable amendment for alkaline soils in terms of cost, availability, ease of handling and long-term efficacy. The variables are the quantity, grain size, and irrigation requirements during reclamation. Pyrite has considerable potential as an amendment of alkaline soils. The application of pyrite on a water-soluble basis is capable of reducing the pH and ESP of the alkaline soil almost as effectively as gypsum. Since gypsum and pyrite not only occur naturally, but are also waste products of some industrial processes, it is possible to get them in large quantities, and relatively cheaply. Before use, they should, however, be checked for the possible presence of some deleterious constituents (such as fluorine in the case of gypsum, and Cd and Ni in the case of pyrite).

The quantity of amendment required depends on (i) the exchangeable sodium content that needs to be replaced, (ii) exchange efficiency, and (iii) depth of soil to be reclaimed.

The quantity of gypsum amendment needed for the reclamation of alkaline soils has to cover both Ca^{2+} required to replace exchangeable Na^+ and that required to neutralize the soluble carbonates. Gupta and Abrol (1990) developed nomograms to facilitate rapid estimation of the gypsum amendment needed for various levels of soil pH and for a broad spectrum of soil characteristics. They recommend the application of gypsum in the form of mixed particle sizes (of less than 2 mm). This allows initial fast dissolution rates at high sodicity, followed by sustained release of calcium over a long period.

The quantity of gypsum amendment (GR, in Mg ha^{-1} or tonnes per hectare) needed, is calculated from the following equation

$$GR = 0.172 \times 104 \times \rho_b \times ds \times n \times (CEC) \, (E_{Nai} - E_{Naf}) \qquad (4.22)$$

where, ρ_b = bulk density (Mg m^{-3}), ds = depth of soil (m), n = Na – Ca exchange efficiency factor (taken as 1.20), CEC = cation exchange capacity (mol$_c$/Mg soil), E_{Nai} and E_{Naf} = initial and desired final fractions of exchangeable sodium.

Graphic approximation of gypsum requirement for illitic soils is given in Fig. 4.4. Equivalent quantities of different amendments needed can be calculated from the conversion constants given in Table 4.6 (*source:* Gupta and Abrol, 1990, p. 248).

The water requirements for gypsum dissolution and leaching of soluble salts can be estimated from the following equation of Hira et al. (1981):

$$Z^{-1/3} = 1 - K \, E_{Nai} \, (2I/3\rho_b \, D_o \, M_o)^{1/2} \qquad (4.23)$$

where $Z = (m_o - m)/m$, m = amount of gypsum dissolved (g. equivalent), m_o = initial amount of gypsum applied per unit surface of the soil, D_o = initial gypsum particle diameter (cm), ρ_b = bulk density of gypsum (g cm^{-3}), E_{Nai} = initial exchangeable Na per unit surface of soil (g. equivalent), I = depth of irrigation water (cm), and K = empirical constant (cm^{-2}).

Hira et al. (1981) calculated that 6.8 cm of irrigation water would be needed to dissolve 99% gypsum of particle size < 0.26 mm. An additional 15 cm of water would be required to leach the soluble salts from 0 to 30 cm soil depth.

The American practice has been to apply 92 to 122 cm of irrigation water to dissolve the amendments. Such high rates of water use is not always feasible. A dissolution equation based on the particle surface area

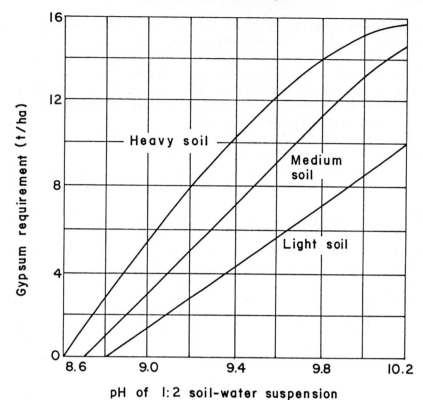

Fig. 4.4 Graphic approximation of gypsum requirements for illitic soils
(*source:* Abrol et al., 1973)

was developed by Hira and Singh (1978). It has been found that by suitable adjustment of the particle size of gypsum, it is possible to use as little as 4 cm of irrigation water to dissolve the amendment.

4.7.8 Reclamation and Management of Salt-affected Soils

The white encrustations on saline soils are generally composed of the chlorides and sulfates of Na, Ca and Mg. Saline soils are reclaimed by reducing the soil salinity to acceptable levels, by leaching. The amount of water needed for the purpose would depend upon the (i) degree of soil salinity that needs to be ameliorated, (ii) quality of irrigation water, (iii) depth to which soil has to be reclaimed, (iv) technique of water application, etc. A rule of thumb is that 80% of the soluble salts from a 50-cm deep soil profile, can be removed by 50 cm of water (Gupta and Abrol, 1990, p. 271). The availability of sturdy, low-cost soil chemical testing equipment has made it possible for the agriculture extension personnel to take prompt action based on in-situ measurements of the soil pH (metallic sensor-based pH meter), salinity (Portable Digital Soil Salinity Tester with a four-electrode probe), and ionic composition (Specific Ion Analyzer).

Rice can be grown in a variety of agroclimatic conditions. Rice culture and incorporation of organics are effective in the reclamation of alkaline soils. They increase the rhizosphere CO_2 concentration, which in turn brings down the soil pH and exchangeable sodium content through the mobilization of the soil calcium carbonate.

Agroforestry constitutes a useful land-use option for degraded soils. Trees, such as *Prosopis juliflora, Acacia nilotica,* and *Casuarina equisetifolia,* and grasses, such as *Diplachne fusca* (Karnal grass), *Brachiaria mutica* (Para grass) and *Chloris gayana* (Rhodes grass) tolerate alkaline soils and ameliorate them.

Plant growth in saline soils is inhibited due to the inherently low level of nitrogen, high exchangeable sodium, and low water solubility of plant nutrient elements as a consequence of high pH and $CaCO_3$. Experience has shown that the nutritional status of the alkaline soils may remain subnormal, even after they have been properly amended. There was a reduction in the efficiency of the N fertilizers because of the volatilization losses of ammonia in the alkaline environment. An effective way to improve the fertilizer efficiency is the pre-puddling application of urea fertilizer.

CHAPTER

5

Climate Change Impacts on Soils

There is no such thing as a free lunch
— Barry Commoner "Laws" of Ecology

5.1 WHAT IS CLIMATE CHANGE?

This chapter would deal with three issues related to climate change: (i) causes and dimensions, (ii) likely impact on soil-based ecosystems, and (iii) possible strategies to mitigate its adverse consequences with respect to soils. It draws extensively from the massive document compiled by the Intergovernmental Panel on Climate Change (IPCC, 1996).

Fossil fuels (coal, oil, and natural gas) account for 88% of the world's commercial primary energy. Burning of fossil fuels leads to the production of climate-relevant emissions of carbon dioxide (CO_2), methane (CH_4), nitrogen oxides (NO_x), carbon monoxide (CO), and volatile organic compounds (VOC). Atmospheric CO_2 concentrations have risen from a base of 280 ppmv (parts per million, volume) before the Industrial Revolution to 370 ppmv at present; they are projected to rise to 560 ppmv ($2 \times CO_2$) by 2065. These emissions cause climate change, particularly in the form of anthropogenic greenhouse effect, or global warming as it is popularly known.

The extent of contribution of emissions from various sources to the greenhouse effect is as follows:
  (i) Energy industries: about 50%, of which CO_2 accounts for 40%,
  (ii) Chemical products, particularly CFCs: about 20%,
 (iii) Destruction of tropical rain forests and related causes: about 15%,
 (iv) Agriculture and others (e.g. waste deposit sites, etc.): about 15%.

The most compelling evidence for recent global warming is found in the rapidly melting icecaps in the alpine regions of the tropics and subtropics, such as those of the Peruvian Andes and eastern Tibet. The Himalayas have about 5000 glaciers, but only a small number of them have been subjected to regular monitoring. The 5-km long Dokriani Bamak glacier in Garhwal Himalayas had retreated by 20 m in 1997, the highest annual retreat thus far.

Under the caption, "Adding insult to injury", *Eos* (a journal of the American Geophysical Union) of May 20, 1997, reported the conclusions of the economic analysis of the global climate change. The analysis was made by Michael Schlesinger, an atmospheric scientist of the University of Illinois, assisted by economists from Yale University, and covered agriculture, forestry, coastal resources, energy, and tourism. An increase of 2° C in the global mean temperature, which is expected to occur by 2050, would benefit the developed countries to the tune of USD 82 billion per year, whereas the developing countries would stand to lose USD 40 billion per year! This is so because most of the industrialized countries happen to be characterized by temperate and cold climates. A 2°C rise in temperature in these regions will have a number of beneficial consequences, such as: (i) the areas in which crops can be grown will enlarge substantially, (ii) the growing season will be extended (say, by about three weeks), making it possible to grow crops of longer duration, and (iii) the energy required for home heating will be reduced (say, by about 15%).

Thus, the industrialized countries, the principal emitters of greenhouse gases, and therefore the perpetrators of "environmental crime", are going to be the gainers, whereas the developing countries, which are the victims, stand to lose. Among the developing countries, the island nations are expected to suffer the greatest economic losses from global warming. These countries have long coastlines, tend to be strongly dependent on tourism, and have small, undeveloped economies. For such countries, global warming is no longer an academic issue. Most of them stand to lose a good chunk of their territories, and some islands may even disappear altogether. When a western delegate, in the course of UN deliberations, remarked that global warming has not been conclusively proved, Ambassador Robert F. Van Lierop, Permanent Representative of the UN from Vanuatu (a small island in the Pacific), is reported to have retorted, *"The proof, we fear, will kill us"*. By the time global warming is proved beyond a shadow of doubt, his country probably may not be around!

5.2 CLIMATE CHANGE MODELS

Our ability to predict future climatic change is critically dependent upon our understanding of the specific causes for past and present changes. We must bear in mind the fact that climate change is influenced not only by greenhouse gases, but also by changes in solar irradiance and major volcanic eruptions. On the basis of the analyses of Northern Hemisphere temperature variations over the past 600 years, it was found that (i) up to the start of this century at least, changes in recorded temperatures are more closely linked to changes in solar irradiance than to either volcanic or greenhouse gas variations, and (ii) but, during this century, increases in greenhouse gas emerged as the dominant forcing mechanism. Three of the past 8 years were the warmest across the Northern Hemisphere since at least 1400 A.D. (Mann et al., 1998).

An examination of the trends in temperature and sea-level pressure anomalies since 1976 and especially during the 1990s, is suggestive of the possibility that the frequency and intensity of ENSO (El Niño Southern Oscillation) events may be a response to global warming. This theory is being checked through a study of the "natural archives", such as, tree rings, annually banded corals and ice cores from tropical glaciers, and the record of flood events preserved in lake sediments.

Two key facts emerge from the latest drillings in Antarctica: firstly, the unprecedented rise in atmospheric methane and carbon dioxide concentrations during this century are attributable to human activities, and secondly, there is close coherence between the changes in temperature and the greenhouse gases during the last 400,000 years. The coherence is particularly marked during the periods of rapid warming at the end of each period of glaciation. Intensive research is ongoing to reconstruct as quantitatively as possible, the sources, sinks, and fluxes of carbon within the Earth system as a whole during periods of rapid change, on the basis of the records preserved in ice and sediment cores.

The Intergovernmental Panel on Climate Change (IPCC, 1996) generated five climate models taking into account greenhouse gases and aerosols. They predict that by 2100 the mean surface temperature could rise 1 to 3.5° C, depending on the assumed factor of climate sensitivity (defined as the equilibrium change in annual surface temperature due to doubling of atmospheric concentrations of CO_2 or equivalent doubling of other greenhouse gases). Some non-IPCC models, however, project a much higher global temperature rise of 3 to 9° C, most probably by 5° C, by 2100. Generally, high latitudes are projected to warm more than the global average.

IPCC models (1996, p. 22) suggest the following consequences of climate change:

1. Climate warming will enhance evaporation. There will be an increase in global mean precipitation and the frequency of intense rainfall. These consequences are, however, unlikely to be universal. Some land regions may not experience an increase in precipitation, and even those that do may experience decreases in soil moisture because of enhanced evaporation. There may be increased precipitation at high latitudes during winter and a decrease in soil moisture in midlatitudes during summer.

2. Severe droughts and floods are likely to occur in some places. It is unclear what effect the climate change would have on the frequency of extreme weather events, such as tropical storms, cyclones, and tornadoes.
3. The climate change will have potential adverse effects on physical and ecological systems, human health, economy, and quality of life.
4. Gerstle (1992) gave the following estimates of sea-level rise in the next century:

Estimate	Year 2050	Year 2100
High	0.5 m	1.6 m
Low	0.1 m	0.4 m
Median	0.3 m	0.8 m

5.3 EFFECTS OF CLIMATE CHANGE ON SOIL PROCESSES AND PROPERTIES

Rounsevell and Loveland (1994) and Tinker and Ingram (1994) studied the effects of climate change on soil processes and properties.

Soil is formed through the interaction of several variables, notably parent material, climate, organisms, relief, and time. By affecting a number of crucial soil processes, climate change will have a bearing on the ability of the soil to support particular natural or agricultural communities. Future distribution of fauna and flora may possibly result in new combinations of soils and vegetation. There is likely to be a mismatch between vegetation and soils and it is not known what effect this will have on the functioning of the ecosystems.

Soil processes that will respond most rapidly (over periods of months or years) and are likely to have the greatest effect on ecosystem functioning, relate principally to changes in the soil water regime and turnover of organic matter and the related mineralization and immobilization of nitrogen and other nutrients. Rise in temperature will only marginally affect inorganic reactions such as ion exchange, adsorption, and desorption. A change in the soil moisture content, however, could significantly affect the rates of diffusion and thus the supply of mineral nutrients such as P and K to plants. This could well alter the species composition of plants in natural systems and may require adjustments to nutrient management and fertilizer use in agriculture.

Soil Carbon Dynamics

According to the estimate of Eswaran et al. (1993), the global pool of soil organic matter is 1500 Gt of C (G = giga = 10^9), distributed as follows: 600 – 700 Gt C in aboveground biomass of vegetation and 800 Gt C in the atmosphere. The oceans contain 40,000 Gt C, which is not directly related to the soils.

Though peatlands (such as those in Alaska, Canada, Scandinavia, and the former Soviet Union) make up just 2.5% of the earth's surface, they contain about one-third of the soil carbon pool. Investigations are in progress to determine whether peatlands will continue to function as a global sink for carbon or will become a major source of greenhouse gases. If, for instance, a situation arises whereby carbon in peatlands could get converted to carbon dioxide or methane, this could have a profound impact on global warming. It is possible, however, that vegetation would change in response to the ambient temperature and moisture, and this could affect the decomposition rates, nutrient availability, and the oxidizability of carbon (*Eos*, Sept. 16, 1997).

Most carbon in the soils is associated with organic matter. However, carbonate-C in calcareous soils, and charcoal in ecosystems subject to frequent fires, may be important in some situations. The soil type, land-use, and climate have a bearing on the amount of organic matter that a soil could contain. Climate plays an important role in the release and sequestration of CO_2.

Most of the organic matter reaching the soil is respired by the soil organisms within a few years, at rates which are dependent upon the climatic conditions and degree of decomposability of the litter. Some highly resistant material may remain unchanged in the soil for hundreds or even thousands of years, even if the ambient physicochemical conditions change greatly.

The ecosystem processes are partly driven by external factors such as solar radiation, temperature, precipitation, air humidity, and atmospheric CO_2 concentrations. The Net Primary Production (NPP), which is the increase in plant biomass of carbon per unit of landscape, is equivalent to gross primary production (all carbon fixed through photosynthesis) minus plant respiration. NPP will increase from 1% per °C increase in temperature in ecosystems with a mean annual temperature of 30°C, to 10% per °C at 0°C (Leith, 1973).

Climate change has a bearing on soil water availability. The latter is related to the ratio of precipitation to potential evapotranspiration, or the ratio of actual to potential evapotranspiration. Many ecosystems experience shortage of soil water (drought) during part or most of the year, which limits their potential carbon gain. Thus, soil water availability can be affected by changes in either gains (precipitation) or losses (evapotranspiration) of water as a consequence of climate change.

The climate change may bring about an increase in the amount and intensity of rainfall in some regions. This would largely affect the surface runoff and partly influence infiltration through the soil. Water infiltration rates are dependent on the characteristics of the soil, such as soil texture and structure, slope, vegetal cover, soil surface roughness, surface crusting, and land management. Infiltration is also controlled by soil water content because saturated soils are unable to absorb any more water, and very dry soils would take a long time to become rewetted. Infiltration takes place most effectively in the case of aggregated soils with good structure, and aggregation is strongly depending upon the organic content of the top soil.

Soil organic matter content in conjunction with particle size distribution, bulk density, and soil structure would determine the water-retention capacity of a soil. Thus, any decrease in the quantity of soil organic matter as a result of (say) faster decomposition could reduce infiltration rates and soil water retention and thereby accentuate water stress to which a plant may be subjected.

Changes in climatic conditions and land-use generally affect both NPP (Net Primary Production) and the rate of decomposition of organic matter. If the rate of increase of NPP is greater than the rate of decomposition of organic matter, there will be an increase in soil organic matter. Per contra, the soil organic matter would decrease if the decomposition rate exceeds NPP.

As could be expected, the greater the availability of water, the higher the rate of production of soil organic matter. At a given level of water availability, the production of organic matter increases with decreasing temperature. Soil moisture and temperature have a profound effect on all processes involving microbes. Increasing temperature would enhance both NPP and the decomposition of organic matter, the latter more so. This situation has a profound consequence — although global warming would increase global NPP, it could still happen that actual soil carbon storage may decrease and more CO_2 may be added to the atmosphere.

NPP has a bearing on the mineralization of nitrogen and phosphorus. If the greater availability of NPP leads to the enhanced immobilization of nitrogen and phosphorus in the soil organic matter, NPP will decrease, but the soil organic matter will increase only marginally. On the other hand, there is evidence to show that increasing CO_2 concentrations promote nitrogen fixation and possibly uptake of mycorrhizal phosphorus, and thereby lead to significant increase in soil organic matter.

Tillage practices determine the retention of carbon in the soil; the less the tillage, the greater the carbon retention.

Knowles (1993) gave an excellent account of how the processes of production and consumption of methane influence Global Climate Change. The total emissions of methane are of the order of 535 ± 75 Mt CH_4 y^{-1}. The following are the main sources of soil methane (in terms of Mt y^{-1}): (i) natural wetland soils: 55 – 150, (ii) microbial degradation of organic substrata in paddy (rice) soils: 20 – 100, (iiii) landfills: 20 – 70, and (iv) termites: 10 – 50.

Soils may serve as sources or sinks for methane, depending on the relative magnitudes of genesis and consumption of the gas. Thus, while more than half of the total emissions of methane are associated with soils, they serve as sinks to the extent of 30 ± 15 Mt of CH_4 y^{-1}. All processes involving methane have a strong biological mediation and are influenced by variables such as organic substratum supply, temperature, hydrologic conditions, pH, Eh, aeration and salinity — all of which are involved in climate change directly or indirectly. Increasing temperatures may affect methane fluxes, either by bringing about changes in the rate of methane formation in lakes and wetlands, or by altering the relative rates of methane synthesis and oxidation.

Soil Nitrogen Dynamics

Bradbury and Powlson (1994) discussed the potential impact of global environmental change on the nitrogen dynamics of arable soils.

There are several sources for nitrogen in the soil: (i) addition as inorganic fertilizer or organic manure, (ii) wet or dry deposition from the atmosphere, and (iii) transfer from the soil through biological nitrogen fixation. Nitrogen is also released (mineralized) by the microbial decomposition of soil organic matter. Mineral nitrogen may be taken up by plants or reabsorbed by soil microorganisms; it may be leached as nitrate to ground and surface waters or emitted to the atmosphere in gaseous forms after nitrification, denitrification, or volatilization of NH_3. These processes are strongly influenced by temperature, soil moisture, plant characteristics, and indirectly, by atmospheric CO_2 concentrations.

Natural soils emit about 6 Mt N yr^{-1}, and cultivated soils 3.5 Mt N yr^{-1}. Soils contribute more than half of the total N_2O emitted to the atmosphere. The following factors influence the biological processes responsible for N-emissions from the soils: (i) environmental variables, such as soil temperature, moisture content, and aeration status, and (ii) agricultural management practices, such as fertilizer regime, cultivation method, and cropping systems. Denitrification rates increase with increasing soil water content.

Levels of nitrogen and sulfur deposition are also increasing in many parts of the world through inputs from industrial pollution and agricultural activities. This may be having beneficial effects on NPP of many ecosystems in the short run, but may have deleterious consequences in the long run due to nutrient imbalances and further acidification of the soils.

Although climate change may lead to small changes in the nutrient capital of the soils or the ratio of carbon to the nutrients, large-scale changes are unlikely in the short term (decades).

Soil Biodiversity

It is likely that changes in soil moisture and temperature arising from climate change will affect the abundance of species within the soil microbial and faunal populations (Arnolds and Jansen, 1992), though the magnitude of such changes may be much smaller than those caused by changes in land-use. At this stage, it is unclear as to what extent biodiversity will be affected by these developments.

The increase in concentrations of atmospheric CO_2 may lead to two consequences: the composition of organic carbon compounds entering the soil from roots and root exudates may change and their quantity may increase. Consequently, the species composition of the rhizosphere population may get modified, which, in turn, could affect the ability of soil-borne pathogens to infect plant roots. We are not sure at this stage whether this is a good thing to happen or not. Possibly, the soil organisms and soil fauna would adjust to the emerging climatic conditions slowly and, consequently, might not be able to keep pace with the rate of migration of the higher plants.

5.4   CLIMATE CHANGE AND COASTAL LOWLANDS

Miura et al. (1994) gave a detailed account of the development and management of lowlands, including the impact of sea-evel rise on lowlands.

The term, "lowland" covers a broad spectrum of lands affected by fluctuating water levels. Coastal areas contain a large percentage of lowlands, variously described as swamps, wetlands, tidelands, mangrove forests, etc. Coastal wetlands play a valuable role in flood protection and fisheries production.

Coastal areas tend to be more densely populated than inland areas, due to economic reasons: ports and shipping, fisheries, tourism, existence of large cities and location of industries, such as oil refineries, etc. The great economic importance of these areas arises from the projected estimate that by circa 2025, about 70% of the world's population will be living within 50 km of the coastline. This would lead to the development of megacities (i.e., those with a population of more than 10 million) in the coastal areas. The total number of megacities in the world rose from 3 in 1972 to 13 in 1992. The trend in this regard is far more pronounced in the case of the developing countries (one megacity in 1972 versus 9 megacities in 1992), and the trend is continuing.

In order to understand how the sea-level rise would affect the coastal wetlands, we must understand the basic dynamics of the coastal wetland ecosystems. Rising sea level affects the hydrology, hydrodynamics, and sediment dynamics of the coastal zone. Due to geological and other reasons, some coasts may be rising, and some coasts may be subsiding at a given point of time.

The total area of lowlands in the world is about 180 Mha. The following countries have more than one Mha of lowlands (the figures in parentheses are in terms of Mha):

Africa: Egypt (1.8)

Asia and Oceania: Bangladesh (8.5), China (5.6), India (4.4), Indonesia (43.0), Japan (2.6), Korea (1.6), Vietnam (1.1),

Europe: C.I.S. (1.4), Italy (1.3), Netherlands (2.0),

North America: USA (88),

South America: Argentina (3.5), Brazil (5.5)

The coastal lowlands are subject to many natural hazards, such as tides, floods, cyclones, storm surges, etc. These events arise from the coupling between oceanic and atmospheric phenomena and coastal topography. The geomorphic, pedogenic, and hydrologic elements in the evolution of coastal lowlands (as a consequence of sea-level rise and possible modifications in the frequency and intensity of storms and tides), are so closely interlinked as to be inseparable. For instance, saline incursions as a consequence of sea-level rise would not only salinize the soils, but also degrade the groundwater resources.

Sea-level rise may be affected by a long-term trend (which may be of the order of 2 mm y^{-1}) and short-term trend for a few years (which may be of the order of 10 mm y^{-1}). El Niño phenomena may cause a sea-level rise of about 20 mm y^{-1}.

The change in rise of the Relative Sea Level (RSL) corresponds to the net effect of the global sea-level rise and the vertical displacements of the coastal zone.

A self-regulating mechanism operates with regard to the coastal wetlands because the ecosystem can respond in ways that nullify in part the consequences of outside forcing. When a wetland elevation is low relative to sea-level, inundation by tides brings in a supply of sediments and nutrients to the wetland. This in turn promotes the growth of vegetation, which then traps the sediment more effectively. Accretion is more directly affected by the rate at which organic matter is produced by the wetland. The elevation of wetland rises, making tidal inundations less frequent.

Coastal wetlands are subjected to a number of stresses, such as herbivory and storm events. Sometimes a storm event may change the ecosystem so drastically as to be beyond the innate capacity of the wetland to regulate itself. River control and dredging practices should be carefully managed so as not to degrade the wetlands.

Four modes of wetland response to rising sea level may be considered, depending on the magnitudes of the landscape slope and the sediment supply (Table 5.1 and Fig. 5. 1; *source:* Brinson et al., 1995):

Table 5.1 Modes of wetland response to rising sea level

Condition	Wetland response
(a) High sediment supply and gentle slope	The availability of suspended sediment enables the wetland to expand laterally into open water areas, even as wetland accretion, in response to rising sea level, drives the migration of wetland inland over gently sloping, upland topography
(b) Low sediment supply and gentle slope	In a sediment-poor coastal setting, the estuarine boundary retreats due to erosion, and wetland shifts landward over time
(c) High sediment supply, and steep slope	The migration of wetland's terrestrial boundary is stalled due to a steeply sloping upland topography or a man-made structure like a dike
(d) Low sediment supply and steep slope	Similar to (c), but erosion occurs

SEDIMENT SUPPLY

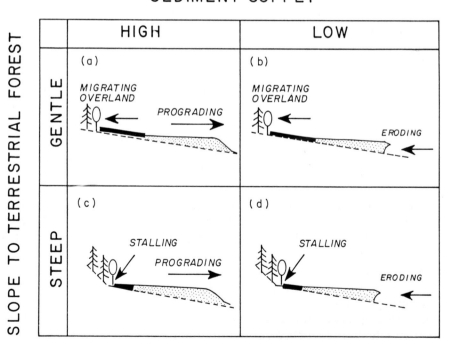

Fig. 5.1 Modes of wetland response to rising sea level
(*source:* Brinson et al., 1995 © Estuaries)

5.5 CLIMATE CHANGE AND SOIL SALINIZATION

The rise in sea level as a consequence of global warming will increase tidal ingression through creeks along the coasts. This will aggravate the salinity of the soils along the coast and damage the fragile wetland ecosystems. Expected losses in soil productivity would depend upon the level of sodicity.

The impact of climate change on soil salinization can be illustrated with the example of India (Sehgal and Abrol, 1994).

In India, saline soils occur mostly along the coastal areas and some inland irrigated areas. Soil salinity affects 10 Mha of land in India, including 2.5 Mha in the Indo-Gangetic alluvial plains. Sea-level rise would have an adverse affect on the 20 Mha coastal ecosystems along the 7,000-km long coastal belt. For instance, a 70 cm rise of sealevel in the twenty-first century would inundate 25% of the coastal areas of Kerala and Tamil Nadu and the vast deltaic areas of the Sunderbans in West Bengal, etc. Enhanced salinity due to salt-water incursion along the west and east coasts would reduce soil productivity, depending on the sodicity level and the type of soil, as follows:

Exchangeable sodium percentage (%)	Loss (%) of productivity in alluvium-derived soils (fluvisols)	Loss (%) of productivity in black soils (vertisols)
up to 5	nil	up to 10
5 – 15	< 10	10 – 25
15 – 40	10 – 25	25 – 50
> 40	25 – 50	> 50

This indicates that (i) the higher the exchangeable sodium percentage, the greater the productivity loss, and (ii) vertisols would be more affected than fluvisols.

5.6 CLIMATE CHANGE AND DESERTIFICATION

The magnitude of land degradation caused by soil erosion by water (Table 5.2) and soil erosion by wind (Table 5.3), has to be examined in the context of regional distribution of drylands in the world (Table 5.4), causes of desertification (in terms of percent of desertified land) (Table 5.5), and the extent and severity of desertification (Table 5.6) (Hulme and Kelly, 1993).

Land degradation has an adverse effect on the physical, chemical, and biological quality of land and its productivity. It is already a problem in many developing countries, particularly of sub-Saharan Africa. According to Falkenmark et al. (1990), drought occurs when the available moisture in the root zone falls below 30% of the water-holding capacity of the soil.

Table 5.2 Soil erosion by water (current rates and future trends)
(*source* : IPCC, 1996)

Region	Area at risk (M ha)	Denudation rate (mm y^{-1})	Dissolved load (Mt y^{-1})	Future trends
Africa	227	0.023	201	+
Asia	441	0.153	1592	+
South America	123	0.067	603	+
North & Central America	106	0.055	758	+ (Centr. Am) + /– (North America)
Europe	114	0.032	425	+/–
Oceania	83	0.390	293	+
World	1094	0.079	3872	+

+ = increased risk ; – = decreased risk

Table 5.3 Soil erosion by wind (*source:* IPCC, 1996)

Region	Area at risk (Mha)	Future trends
Africa	186	+
Asia	222	+
South America	42	–
North & Central America	40	–
Europe	42	+/–
Oceania	16	+
World	548	+

+ = increased risk ; – = decreased risk

Table 5.4 Regional distribution of dry land in the world (in 10^3 km^2)
(*source:* IPCC, 1996)

Zone	Africa	Asia	Austra-lasia	Europe	North Amer.	South Amer.	Total
Cold	0.0	1082.5	0.0	27.9	616.9	37.7	1765.0
Humid	1007.6	1224.3	218.9	622.9	838.5	1188.1	5100.4
Dry, sub-humid	268.7	352.7	51.3	183.5	231.5	207.0	1294.7
Semi arid	513.8	693.4	309.0	105.2	419.4	264.5	2305.3
Arid	503.5	625.7	303.0	11.0	81.5	44.5	1569.2
Hyper arid	672.0	277.3	0.0	0.0	3.1	25.7	978.1
Total	2965.6	4255.9	882.2	950.5	2190.9	1767.5	13012.7

Table 5.5 Causes of desertification in terms of percent of desertified land (*source:* IPCC, 1996)

Regions/ countries	Over culti-vation	Over-stocking	Fuel and wood	Saliniza-tion	Urbani-zation	Others
(1)	45	16	18	2	3	16
(2)	50	26	21	2	1	—
(3)	25	65	10	—	—	—
(4)	10	62	—	9	10	9
(5)	22	73	—	5	—	—
(6)	20	75	—	2	1	2

(1) Northwest China, (2) North Africa and Near East, (3) Sahel and East Africa, (4) Middle Asia, (5) United States, (6) Australia

Table 5.6 Extent and severity of desertification
(*source* : IPCC, 1996)(under each category, desertified area is expressed in 10^3 km^2, and as % of the total dry lands ; dry lands = arid + semiarid + dry subhumid)

Region	Light	Moderate	Strong	Severe
Africa	1180 (9%)	1272 (10%)	707 (5.0%)	35 (0.2%)
Asia	1567 (9%)	1701 (10%)	430 (3.0%)	5 (0.1%)
Australasia	836 (13%)	24 (4%)	11 (0.2 %)	4 (0.1%)
N. America	134 (2%)	588 (8%)	73 (0.1 %)	0 (0.0%)
S. America	418 (8%)	311 (6%)	62 (1.2 %)	0 (0.0%)
Total	4273 (8%)	4703 (9%)	1301 (2.5 %)	75 (0.1%)

Climate change models predict that in some regions there will be increase in drought, with rainfall of higher intensity and more irregular distribution. This could lead to greater land degradation, manifesting itself in the form of "loss of organic matter and nutrients, weakening of soil structure, decline in soil stability, and increase in soil erosion and salinization " (IPCC, 1996, p. 29).

Some areas may benefit from increased rainfall and soil moisture but they may experience stronger leaching, leading to increased acidification and loss of nutrients. This can be mitigated by the addition of lime and fertilizers and the adoption of appropriate conservation practices.

Barring a few exceptions, deserts are likely to become hotter but not significantly wetter. Organisms that now exist near their heat-tolerance limits will be adversely affected by temperature rises. There is uncertainty as to how climate change will affect the water balance, hydrology, and vegetation in desert regions, though it is safe to say that the effects would vary significantly among the regions (IPCC, 1996, pp. 28-29).

5.7   CLIMATE CHANGE AND THE BIOMES

An ecosystem has abiotic and biotic components which interact in complex ways. For instance, the plant and animal communities living in a soil modify it. The process of equilibration between them may take decades or even longer. All organisms have characteristic "fundamental niches" or environmental conditions (such as, range of temperature, precipitation, and soil conditions) in which they can potentially grow and reproduce. Climate change may bring about a situation whereby an organism may find itself outside its fundamental niche and therefore unable to grow and reproduce.

A biome is "a grouping of similar plant and animal communities into broad landscape units that occur under similar environmental conditions" (IPCC, 1996, p. 864). On the basis of data regarding precipitation, temperature, humidity, and windspeed, Neilson and Marks (1994) developed a model (MAPSS: Mapped Atmosphere-Plant-Soil Systém) which can indicate the natural vegetation that can be supported at a given site. They identified six major biomes in the world: (i) Tundra, taiga-tundra and ice; (ii) Boreal forests; (iii) Temperate forests; (iv) Tropical forests; (v) Savannas, Dry forests and Woodlands; (vi) Grasslands, Shrublands, Deserts. They plotted the potential distribution of these biomes under the current climate conditions, as also the projected distribution of the same biomes to simulate the effects of doubling of carbon dioxide concentrations (IPCC, 1996, p. 27).

Figure 5.2 shows how the distribution of major biomes is related to the mean annual temperature and mean annual precipitation. It should be pointed out, however, that the distribution of biomes is also dependent upon seasonal factors (such as length of the dry season and lowest absolute minimum temperatures), soil properties (such as water-holding capacity) and land-use history (such as agriculture, grazing, and disturbances like fire).

How climate change will affect the vegetation zones in a given physical setting may be illustrated with the case of a mountain site. Mountains cover about 20% of the continental surfaces. Most of the major rivers in the world (such as the Ganges and Amazon) originate in the mountains. Mountain ecosystems are under considerable stress from humans. It is expected that global warming will cause the distribution of vegetation to shift to higher elevations, with consequent changes in ecosystems. Some species may disappear. Figure 5.3 gives the distribution of vegetation zones at a hypothetical dry temperature mountain site in two situations: current climatic situation, and a situation wherein the temperature will go up by 3.5°C and precipitation by 10%.

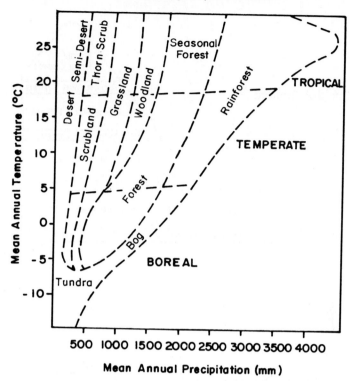

Fig. 5.2 Distribution of major biomes in relation to temperature and precipitation (*source:* IPCC, 1996, p.70, © IPCC)

A warmer climate will reduce the volume and extent of glaciers, and the extent of permafrost and seasonal snow cover. This will affect soil moisture and soil stability, and thus will have an impact on a variety of socioeconomic activities (e.g. agriculture, tourism, hydropower, and logging).

5.8 COPING WITH CLIMATE CHANGE

IPCC (1996, pp. 180-185) recommends a number of eminently practical strategies for mitigating the adverse consequences of climate change. These need to be adapted in the context of individual countries and regions. It is possible to design microenterprises whereby environmentally desirable consequences can be realized as spin-off benefits of economically viable activities.

Water-use and Management

Though the water availability in soil in an area is directly related to the magnitude and distribution of precipitation in that area, there

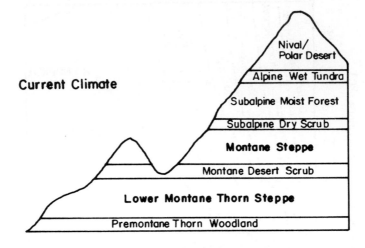

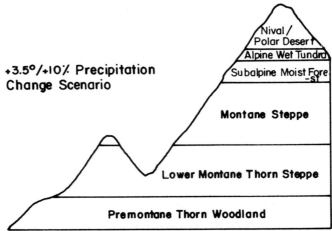

Fig. 5.3   Projected changes in the distribution of vegetation zones at a hypothetic mountain site
(*source:* Beniston, 1994, © Routledge)

may be large differences (even by a factor of ten) in water availability at different soil sites, depending on topography, geomorphic setting, characteristics and depth of soil. A number of techniques are available to conserve water in the soil: tillage and mulching practices, runoff farming, water harvesting, and wadi diversion. None of these techniques are new; they were known and practiced in western China, Iran, Jordan, northern Africa, Spain, etc. for millennia. In some of these countries, these practices have gone out of fashion. Wisdom lies in restoring these practices not only in the lands where they were practiced earlier, but also in other lands where they are suitable.

Where water is available for irrigation, conveyance and application losses can be minimized by new techniques such as drip irrigation and underground irrigation. As much as 50% of the water can be conserved by these techniques.

Biotechnology can be used to develop water-efficient strains of crops. Preference should be given to low-water need crops — for instance, as against 300 ha. cm of water needed for the irrigation of one ha of sugarcane, the irrigation needs per hectare of sorghum and pearl millet are as low as 20 ha. cm and 10 ha. cm respectively. Crops growing during winter should be preferred, as they would need less water, since PET (potential evapotranspiration) is much less in winter than in summer.

Though drainage water generally tends to have a high salt content, it can still be reused to grow salt-tolerant crops: cash crops (such as asparagus and sugarbeet), fodder crops (such as saltbushes, fleshy sainfoin, tall fescue and strawberry clover), and tree crops for timber. Care has to be taken, however, to ensure that irrigation with salty drainage water does not salinize the soil in the long run.

Land-use systems

Farmers could adopt a two-pronged drought-enduring strategy whereby (i) they reduce the stock to a level which would allow the ecosystem to recover after a drought, and (ii) they grow fodder shrubs and trees (e.g. saltbrush plantation, spineless fodder cacti, wattles, mesquites, and acacias) in strategic locations, in order to provide an extra source of fodder to the animals when drought occurs.

Another good strategy is game ranching. Some animals (such as some species of Cervidae, antelopes, and ostriches) are better adapted to dry conditions than cattle; these could be raised in ranches. Presently, there are more than 20 Mha of game ranches in South Africa, Namibia, Botswana, and Zimbabwe. When coupled with commercial hunting and green tourism, such game ranches can be highly profitable economic ventures.

Some pastoralist tribes in East Africa (such as Somali, Rendille, Gabra, Samburu) have switched from cattle husbandry to camel-rearing, and some others (e.g. Turkana) find it more convenient to raise goats rather than sheep.

Saline Agriculture

The world has about 180 Mha of lowlands, some of which are faced with problems of soil salinity. The rise in sea level as a consequence of global warming will mean increased sea-water incursion into the coastal lowlands. This should give more impetus for developing saline agriculture (Le Houerou, 1986).

Figure 5.4 shows the growth response of crops to soil salinity (expressed in deci-Siemens per meter — dS m^{-1}). Many halophytes (e.g. *Suaeda maritima*) have increased yields at low salinity levels. Salt-tolerant crops (e.g. barley) maintain yields at low salinity levels, but yields decrease as salt levels exceed a certain limit. Salt-sensitive crops (e.g. beans) suffer sharply reduced yields even in the presence of low-level salts.

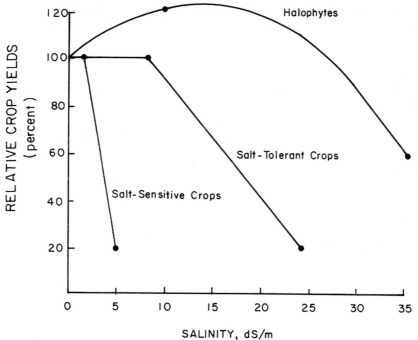

Fig. 5.4   Growth response of the crops to salinity
(*source:* "Saline Agriculture", 1990, p.5, © US Nat. Acad. Press).

The ability of plants to tolerate salt depends upon interactions between the salinity, and various soil, water and climatic conditions. Some halophytes require fresh water for germination and early growth, but can tolerate higher levels of salt in the later stages.

Any strategy for the promotion of saline agriculture has to have the following components:

  (i) Identification of salt-tolerant plants that could serve as food, fuel, fodder and other products, such as essential oils, pharmaceuticals, and fiber;

 (ii) Improving the salt-tolerance of the known plants, for instance, cells of rice (*Oryza sativa*) subject to salt stress and then grown to maturity tend to have progeny with improved salt tolerance (up to 1% salt);

(iii) Exploration for new salt-tolerant wild species that have economic importance, and ways of improving their agronomic qualities.

Some important plants that figure in saline agriculture are as follows:

Food:

Vegetables: asparagus (*A. officianalis*), beet root (*Beta vulgaris*), lettuce (*Lactuca sativa*),

Crops: rice (*Oryza sativa*), barley (*Hirdeum vulgare*), pearl millet or bajra (*Pennisetum typhoides*),

Foliage and leaf protein: Indian saltwort (*Suaeda maritima*).

Fuelwood trees and shrubs:

Prosopis juliflora, Eucalyptus occidentalis, Casuarina obesa.

Kallar grass (*Leptochloa fusca*) can be used for biogas production (energy yield is 15×10^6 kcal/ha).

Fodder:

Kallar grass (*Leptochloa fusca*), silt grass (*Paspalum vaginatum*), salt grass (*Distichlis spicata*), *Atriplex nummularia*

Trees: *Acacia cyclops*

Essential oils:

Ekeda (*Pandanus fascicularis*), *Mentha arvensis*.

Agroforestry

Agroforestry can play a very important role in the development of arid lands and in controlling soil erosion and desertification. It has been used for centuries by rural civilizations in some parts of the world: Kejri [*Prosopis cineraria* (L) Duce] in Rajasthan, western India; *Faidherbia albida* (Del.) Chev. in various parts of intertropical Africa, *Argania spinosa* (L.) Sk. in southwest Morocco, etc. Such agroforestry techniques permit population densities of the order of 60 – 80 people/km^2 even though the rainfall is meagre (< 150 to 300 mm/yr).

The Central Salt and Marine Chemicals Research Institute (CSMCRI), Bhavnagar, Gujarat, India, promotes the development of multipurpose agroforestry plantations to provide economic benefits to the people, while controlling erosion, and bringing semiarid waste lands or saline lands under cultivation. Two examples are given. Jojoba (*Simmondsia chinensis*) is an excellent desert crop. Its seeds yield a valuable industrial oil (45 – 60% in seed), which is used as a lubricant for high-temperature, high-speed machinery, as coating on pharmaceutical tablets, and as a base in cosmetics. Pillu or Kharijal (*Salvadora persica*) is an evergreen, drought-tolerant tree that can be irrigated with saline water. Its seeds yield 40 – 45% oil, which can be used in making soap. Its leaves are used in making sauces and the tender shoots are eaten as salad.

CHAPTER

6

Soils in Relation to Irrigation

Progress is a nice word. But change is its motivator, and change has its enemies

—Robert Kennedy.

6.1 WHY CROP PLANTS NEED WATER

Modern agriculture accounts for 70 – 80% of the use of water. More than a third of the world's harvest is grown on irrigated land. In order to have enough water in the long run, we must develop techniques to use water more efficiently in agriculture, and to keep it cleaner as we do so.

While the world's population has doubled in the past half century, the consumption of meat has quadrupled. This has important implications for grain harvests needed. About 2 kg of grain are needed to produce a weight gain of one kg of chicken. For pork, it is 3 kg and for beef, 8 kg. For the annual global production of 200 million tons of meat, livestock are now fed about 40% of all grain harvested. At a given level of nutrition (say, minimum of 2,200 daily calories), the grain and soybean needed to feed a pig to produce pork will feed a man ten times the number of days, if consumed directly instead of as pork (source: *National Geographic*, Oct. 98).

6.1.1 Role of Water in the Physiology of Crop Plants

The growth and well-being of a crop plant are critically dependent upon the availability of water. Water provides the hydrogen atoms for reducing carbon dioxide in the process of photosynthesis. Nutrients are transported into, within and out of the plants through the medium of water. Water constitutes the basic structural component of plants; more than 90% of plant mass is composed of water.

Water has to be delivered to the plant in tune with the rooting depth of the crop and available water capacity of the soil (150 – 200 mm per meter-depth of the soil for clayey soils, 100 – 150 mm for loamy soils, and 50 – 100 mm for sandy soils).

The rate of water uptake by plants depends upon (i) characteristics of the plant (rooting density, rooting depth, rate of root extension, etc.), (ii) properties of the soil (water retention and conductivity), and (iii) weather conditions (which determine the quantum of transpiration) (Hillel, 1987, p.31).

It was earlier believed that the plant functions are not much affected until the permanent wilting point is reached. Recent studies, however, show that a plant suffers water stress and reduction in growth long before the permanent wilting point is reached. This has to be borne in mind in the application of irrigation.

Crop yields can be maximized only by keeping water readily available to the plant throughout the growing season.

When plants are stressed and transpiration decreases even temporarily, this is manifested in the form of rise in temperature of the crop canopy. Thus water stress can be remotely sensed through infrared radiation thermometers by the measurement of crop canopy temperatures. Since a satellite takes about 90 min to go round the world, frequent and repetitive monitoring of the water stress is feasible. In advanced countries, irrigation decisions (how much, where, when, and how) are taken now a days on the basis of such remotely sensed satellite data.

6.1.2 Crop Evapotranspiration and Water Requirements

Since the water requirement of a crop is conditioned by rate of evapotranspiration in a particular situation, a knowledge of the rate of evapotranspiration (mm d^{-1}) is necessary to plan irrigation.

The rate of evapotranspiration depends both upon meteorological factors (radiation, atmospheric humidity, temperature, wind, etc.) and field factors (wetness and surface properties). Potential Evapotranspiration (PET) corresponds to the "maximal evaporation rate which the atmosphere is capable of extracting from a well-watered field under given climatic conditions" (Hillel, 1987, p. 14). The equations developed by Penman of England in 1948 for determining PET are still used, though with some refinements.

The typical PET values (mm d^{-1}) for different agroclimatic zones are summarized below (quoted from Hillel, 1987, p. 16):

Climatic zones	Mean daily temp. cool (below 20°C)	Mean daily temp. warm (above 20°C)
Tropics and subtropics		
Humid and subhumid	3 – 5	6 – 8
Arid and semiarid	5 – 7	8 – 10
Temperate zones		
Humid and subhumid	2 – 4	5 – 7
Arid and semiarid	3 – 5	6 – 9

MET (Maximal Evapotranspiration) is a useful parameter. It corresponds to maximal seasonal evapotranspiration from a well-watered crop stand of optimal density (Hillel, 1987, p.22). It is usually 0.6 to 0.9 of total seasonal PET. MET is a direct measure of the water requirements of a crop.

The crop water requirements (CWR) are computed from the following equation:

$$CWR = KC \times PET, \tag{6.1}$$

where KC is an empirical crop coefficient and PET is Potential Evapotranspiration.

The KC values depend on climate (higher values for hot, windy, and dry climates, and lower values for cool, calm, and humid climates) and the crop (stage of growth, crop reflectivity, crop height and roughness, degree of ground cover, canopy resistance to transpiration, etc.). For most crops, the KC value for the total growing season varies between 0.6 to 0.9.

Figure 6.1 shows the relationship between crop coeffcent and the stages of growth of an annual crop.

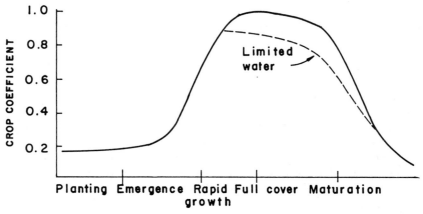

Fig. 6.1   Variation of Crop Coefficient (KC) during the growing season of an
annual crop
(*source:* Hillel, 1987, p. 22, © World Bank).

The method of computing water requirements of a crop is illustrated with the following exercise:

Calculate the water requirements for a crop with four months growing season, and express it as a percent of total PET, given the following particulars:

	May	June	July	August
KC	0.6	0.8	1.1	0.7
PET	5	7	9	9

Water requirement for the crop for the month of May =
KC × PET × no. of days in the month = 0.6 × 5 × 31   = 93 mm
Water requirement for June = 0.8 × 7 × 30   = 168 mm
Water requirement for July = 1.1 × 9 × 31   = 307 mm
Water requirement for August = 0.7 × 9 × 31   = 195 mm
Total water requirement for the season   = 763 mm

Total PET for the season = (5 × 31) + (7 × 30) + (9 × 31) + (9 × 31) = 923 mm
Total water requirement as a percent of total PET = (763/923) × 100 = 83%

It is important to bear in mind that soil water availability does not depend on the soil alone, but on the soil-crop-climate system. For instance, a crop with extensive and dense roots (such as those of small grains) can utilize the soil moisture more effectively than plants with sparse and shallow roots (such as those of potato). A crop growing in arid climates (with consequent high evaporative demand) tends to experience water stress more frequently than the same crop growing in moderate or humid climates.

6.2   WATER-USE–YIELD RELATIONSHIPS

Transpiration is a necessary and productive part of plant activity, but evaporation should be deemed a loss since the plant does not benefit by it. It therefore follows that the evaporation component of evapotranspiration needs to be minimized. This consideration has a bearing on the efficiency of application of an irrigation system. For instance, in sprinkle irrigation, that part of the water which is intercepted by the foliage evaporates rapidly without entering the transpiration system. Similarly, when the whole surface is repeatedly wetted through surface irrigation, there will be avoidable loss of water through evaporation. On the other hand, inadequate irrigation will lead to soil desiccation and water stress. Thus, a balance has to be maintained between the various parameters. Evaporational losses can be minimized through some kinds of irrigation (say, drip irrigation) which wet only a small fraction of the soil surface but do not wet the foliage. This consideration holds good even if the irrigation is frequent.

One would think intuitively that, other things being equal, the more the application of water to the crop, the more its yield. As Figure 6.2 indicates, the actual relationship between the two parameters is not a simple one. When the water supply is not a limiting factor throughout the growing season, a crop plant may be able to transpire at MET rate and attain full potential yield. If water is applied beyond MET needs, it becomes counter-productive and the yield gets reduced.

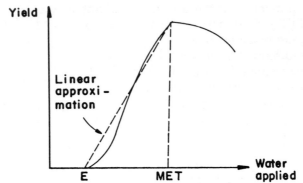

Fig.6.2   Relation of crop yield to water supply
(*source:* Hillel 1987, p. 25 © World Bank)

Doorenbos and Kassam (1979) gave the following relationship between yield and applied water:

$$1 - (Y/Ym) = f\,[1 - (AET/PET)] \qquad (6.2)$$

where Y is the actual yield, Ym is the maximum attainable yield when full water requirements are met, f is an empirical yield response factor, AET is the actual evapotranspiration and PET is the potential evapotranspiration. The values of f for different crops and agroclimatic settings have been reported in the literature. In the above equation, yield refers to dry matter yield. As the grain yield is directly related to dry matter yield, it is safe to state that grain yield depends upon the quantum of applied water, assuming that there are no complications, such as pests and nutrient deficiencies.

Irrigation is, of course, unnecessary when it rains, and the rain is free. But the catch is the uncertainty involved. We can make a projection of the water needs of the crops in time and space, but we cannot know for sure what proportion of this will be taken care of by rain. Statistical models have been developed to address this issue.

6.3   SPAC (SOIL-PLANT-ATMOSHPERE-CONTINUUM)

It is now well recognized that the processes taking place in the soil-plant–water system are interlinked. For instance, the availability of moisture in the soil does not depend upon the characteristics of the soil alone, but is regulated by the interactions between the plant, soil, and climate. It is therefore necessary to study the system in a dynamic and holistic way. Such a system is called SPAC (Soil-plant-Atmosphere-Continuum).

As Hillel (1987, p. 19) put it elegantly, a plant has to live simultaneously in two very different realms, the atmosphere and the soil. The conditions in the two realms vary constantly, but not necessarily in the same manner. A plant therefore has to respond to the sometimes conflicting demands from the two realms. Water is a case in point. A plant has to keep its stomata open in order to absorb carbon dioxide from the atmosphere for the purpose of photosynthesis. The plant loses water, i.e., transpires, through the same stomata. Plants therefore have to extract water from the soil through the root system and store the water. However, if the soil has no moisture for the roots to extract, the plant suffers moisture stress. The plant reacts to moisture stress by partially closing the stomata and reducing the loss of water through them. Transpiration of water cannot be stopped entirely by the closure of stomata, because the leaves continue to lose some water through the cuticular surfaces. This process, however, has an adverse effect on the productivity of the plant. The closure of stomata leads to decreased absorption of carbon dioxide, decreased photosynthetic activity, and ultimately decreased yield of the crop plant. The plant suffers dehydration and wilting when faced with sustained transpiration without sufficient replenishment of moisture.

Figure 6.3 (after Pedro, 1984, as modified by Darnley et al., 1995) illustrates comprehensively the relationship between soils, types of duricrust, thickness of weathering zone, mean annual temperature, mean annual rainfall, Eh, pH, electrical conductivity, and activity of some major cationic and anionic species. It may be noted that prolonged and deep weathering in tropical regions has a profound effect on bedrock chemistry.

The dynamics of the movement of moisture in the soil-plant-atmosphere system are schematically shown in Fig. 6.4.

6.4 MOISTURE CONSERVATION IN DROUGHT-PRONE AREAS

There is much variation in the capacity of the soils to provide water and nutrients to crops, and the needs of the crops for water and nutrients. For instance, one hectare of sugarcane needs 300 ha. cm of water. The same amount of water could irrigate about 8 ha of wheat or 30 ha of pearl millet (known as bajri in India) (Fig. 6.5). Both the supply and demand for water are determined by climate (precipitation, evaporation, transpiration, temperature and radiation, etc.).

6.4.1 Collection and Concentration of Precipitation

Ancient civilizations in parts of Asia and South America have developed agriculture under a variety of adverse conditions — high slopes, rugged terrain, low rainfall, lack of suitable supplies of groundwater, too remote for bringing water through canals, and so on. They could do so because

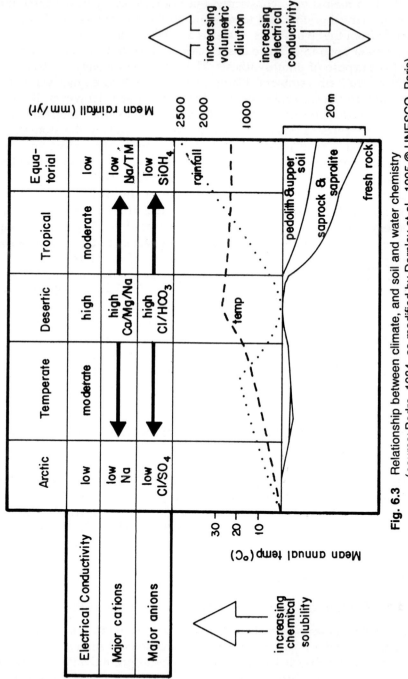

Fig. 6.3 Relationship between climate, and soil and water chemistry
(*source:* Pedro, 1984, as modified by Darnley et al., 1995 © UNESCO, Paris)

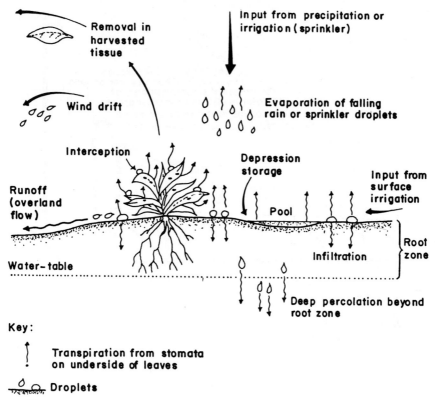

Key:

$\uparrow$ Transpiration from stomata
on underside of leaves

Droplets

Fig. 6.4 Movement of moisture in the soil-plant-atmosphere system
(*source:* Barrow, 1987, p. 82)

they developed simple but efficient techniques of collecting and concentrating precipitation.

For instance, two cm of rain falling over a catchment of 1 ha has a volume of 200 m³. Even if 50% of this precipitation (i.e., 100 m³) is lost due to infiltration and evaporation, it is still possible to save 100 m³ of water if appropriate techniques are used. The water thus stored can be used for irrigation. Any technique of conserving water almost invariably ameliorates soil erosion.

Figure 6.6 shows the following methods of making use of runoff, while at the same time preventing soil erosion: (a) Simple tied ridge and furrow (box ridging), with ties being made with hoe or discplough; (b) Natural catchment and ephemeral stream, with contour bunding to reduce rate of runoff and to guide water to the storage tank(s) — rainfed crops can be grown in the catchment area, and irrigated crops can be grown with water from tank(s); (c) ICRISAT system of broad-beds and grassed furrows/drains — water stored in the tanks can be made use of as

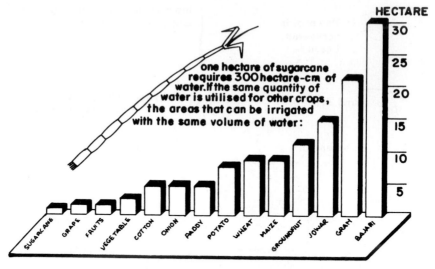

Fig. 6.5 Water needs of different crops
(*source:* Falkenmark et al., 1990 © Univ. of Linköping, Sweden).

supplemental irrigation, or through sprinkler irrigation through bullock-drawn water carts; (d) Crosssection of (c).

6.4.2   Rain- and Storm-water Harvesting

All rainfall does not produce runoff. There is a threshold level of precipitation which is the minimum rainfall needed to produce useful runoff. An area may experience frequent showers but there still may be no runoff because the precipitation may simply evaporate or soak away.

Techniques are available to improve runoff efficiency, which is defined as the yield expressed as a percent of the total precipitation on the catchment. Runoff efficiencies greater than 80% can be achieved by covering the catchment with sprayed silicone compounds, concrete, aluminum foil, butyl rubber sheet, gravel-covered plastic sheet, paraffin wax, asphalt, polyvinylfluoride, etc. Apart from the problem of expense, there is an ever-present risk of the materials being stolen. Besides, plastics are susceptible to damage by wind, sunlight, livestock or pests.

A labour-intensive, and inexpensive treatment is the method of soil grading and compaction. This requires no purchased material and cannot be stolen. It can last for 5-10 years and costs only USD 0.04-0.06 cents/m^2. The cost per 1000 L of yield is USD 0.07-0.19 cents.

Model experiments made by ICRISAT, Hyderabad, India, show that two installments of supplementary irrigation of 20 mm each, will greatly enhance crop productivity. This can be accomplished if one ha of maize field has a water pond of 800 m^3 capacity. Harvested rain water is

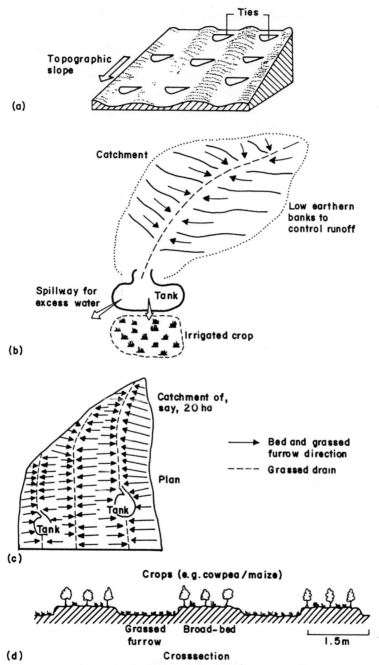

Fig. 6.6 Techniques of conserving runoff
(*source:* Chris Barrow, 1987, p. 172 © Longman's)

recommended to be stored in ponds with limited surface area (to reduce evaporation), and lined with suitable impervious material such as bentonite clay (to reduce loss due to seepage). In the case of deep Alfisols, impervious mixtures of clay, silt, and sodium carbonate, or soil, sand, and cement in the ratio of 4 : 4 : 1, could reduce seepage by 47 to 97%. Spraying of long-chain alkanols and monomolecular films on the surface of water reduces evaporation by 10-35%. It has been reported that emplacement of foamed wax blocks on the water surface is capable of reducing evaporation by 33 to 87%. These techniques are not recommended for use in Africa, as the materials are not readily available locally.

6.4.3   Strategies for Improving the Availability and Use of Moisture

In the case of drought-prone areas (as in, say, sub-Saharan Africa), it is crucially important to improve the availability and use of soil moisture. Four groups of techniques are available for the purpose: (a) improving soil moisture intake; (b) reduction of evaporation losses; (c) reduction of evapotranspiration losses; and (d) reduction of percolation losses. Though some techniques are essentially agronomic, they are included for purposes of comprehensive coverage (Table 6.1).

The following agronomic methods are available to optimize the use of soil moisture: (i) sowing seeds before the main rains break; (ii) maintaining seeding density in inverse proportion to aridity; (iii) fallowing: in some African countries, nitrogen-fixing trees (e.g. *Acacia*) or shrubs (e.g. *Stylosanthes*) are grown during the fallow period; (iv) planting crops in such a manner that they mature one after the other; (v) planting of a crop to go through the previous crop; and (vi) growing two or more crops or two or more varieties of a given crop to spread the risk (so that there will not be total failure of the crop) — for instance, the practice of intercropping groundnuts, maize, and pumpkin in rainfed agriculture in Mozambique, is the most sensible one under the circumstances.

6.4.4   Concurrent Production and Conservation

There are various ways to increase the production of food, fuel, fodder, and fiber. The Green Revolution, based on the use of high-yielding varieties of seeds, and related technical and chemical inputs, did increase the yields of several cereals. The well-off farmer who practiced it derived much benefit from it. But the subsistence farmer working poor soils in arid areas did not share in this prosperity. An alternative strategy was developed in Maharashtra province, India, for this kind of situation

Table 6.1 Strategies for the conservation of moisture (after Barrow, 1987, pp.151-154)

Description of technique	Comments
(a) Improving soil moisture intake	
1. Planting of cover crop(s)—they provide ground cover and reduce runoff.	Inexpensive and easy. Possible benefits are: addition of organic matter to the soil and fixation of nitrogen. Also, the crop may be grazed.
2. Leaving crop stubble/debris (this slows the runoff).	Inexpensive and easy. Reduces erosion. Could add organic matter and nitrogen to the soil. Has grazing potential.
3. Strip cropping—strips of crops, perennials, pasture, etc. arranged roughly along contour.	Inexpensive and easy. Reduces erosion. May have beneficial symbiotic effects, by way of fixing nitrogen, deterring pests, etc. Strongly recommended.
4. Mulches—organic or inorganic material spread on land to slow runoff and reduce evaporation and weed growth.	Mulches appear to be more valuable in increasing infiltration (by slowing the surface flow) than in moisture conservation. Can reduce mudsplash of crop, and weed growth.
5. Tillage—for improving infiltration into poorly permeable soils.	By facilitating infiltration when rains come, tillage speeds planting. But in other situations, can cause erosion and oxidation of organic matter.
6. Chemical treatment—application of "wetting agents" which speed up infiltration.	Easy. Possible pollution risks and costs. Not fully tested.
7. Runoff bunds and terraces—physical structures which retard runoff and promote infiltration.	Inexpensive, involving essentially the labor of the farmer.
(b) Reduction of evaporation losses	
8. Mulches—peat, plant debris, straw, vermiculite, ash, sand, dust, gravel, sawdust, etc.	Reportedly capable of reducing evaporation, but its effectiveness for the purpose is debatable. Reduces rain-splashing and erosion. May deter pests and retard weed growth. Usually applied after the infiltration of moisture. Has to be replaced for each new cropping cycle, as the mulch is ploughed into soil.

(Contd.)

Table 6.1 (*Contd.*)

Description of technique	Comments
9. Antievaporation compounds—these are sprayable emulsions of water and oil or wax, hexadecanol, bitumen, asphalt, latex,etc.	Easy to apply. Binds sandy soils. Reduces erosion due to wind and water. Can be used to build shelter-belts. May be costly or may have pollution risks.
10. Removal of weeds, by the use of herbicides, pulling or burning	Can reduce interception losses. May have pollution risks.
11. Planting of hedges or shelter-belts, in such a manner that they do not take away moisture from shallow-rooted crop plants.	Though these take time to establish, and may have to be protected initially from animals and people, they bring several benefits: reduction of evaporation and wind damage to crops; may increase dew precipitation downwind. May fix nitrogen in the soil; may provide compost for soil improvement; may provide fodder, fuel, and fruits.
12. Windbreaks (structures, built of brushwood, palm fronds or stone barriers).	Reduce evaporation. Protect crops from wind damage.
(c) Reduction of evapotranspiration losses	
13. Cultivating crops with low evapotranspiration.	Such crops have to be identified for a given agroclimatic situation.
14. Removal of deep-rooting weeds from ground crops.	Reduces evapotranspiration. However, may reduce infiltration and increase soil erosion.
15. Spray crops with a suitable compound (such as, kaolinite and water) to reduce the albedo.	Such a spraying reduces solar heating and pest damage. Possible adverse consequences are reduction in yields because of reduction in photosynthesis or respiration of crop. Care should be taken to ensure that the compound sprayed is not toxic, and does not leave a bad taste on the crop material.
16. Spraying of crops with compound(s) such as wax emulsion, silicone, latex emulsion, or apply chemicals which cause crop stomata to close, to retard transpiration.	Same considerations as above
(d) Reduction of percolation losses 17. Addition of hydrophytic material, such as organic matter or the new "Agrosoak" compounds.	Can be applied easily and fast. May be costly.

(Contd.)

Table 6.1 (*Contd.*)

Description of technique	Comments
18. Insertion of an impermeable film, made of plastic, rubber bitumen, below the crop roots.	The film prevents percolation losses. Also, the nutrients are retained, thereby reducing the need for fertilizer application. On the other hand, the insertion may need much labor or special mechanical equipment, and may cause soil drainage problems and salinization.

(Paranjpe et al., 1987; Rao, 1989; Gadgil, 1989, quoted by Falkenmark et al., 1990). The strategy has the following features: (i) controlling soil erosion, maintaining organic matter, moisture and physical properties of the soil, and promoting efficient nutrient cycling; (ii) efficient use of photosynthesis and total stock of water and nutrients in the ecosystem; and (iii) combining perennials with seasonal crops, through integrated tree crop-agriculture and intercropping. The annual precipitation in the area is 500 to 700 mm, with considerable variation between years. Rain water is harvested and stored, and the *assured* water-supply is used to grow *low-water-need food* crops only. In years when there is better than average rainfall, the additional amount of water available is used to grow perennials and cash crops. Preference is given to low-water-need crops, such as maize, sorghum and millet, and special effort is made to preserve soil organic matter. The project has been able to achieve 20 t ha^{-1} (dry weight) in the forestry part, and 6 t ha^{-1} of seasonal crops (sorghum).

6.5 INFILTRATION UNDER VARIOUS IRRIGATION REGIMES

6.5.1 What is Infiltration?

In filtration (or surface intake) refers to the rate of entry of water into the soil under the action of gravity. It is greatest when the land is dry and takes place either at the start of precipitation or the earliest irrigation application. The rate of infiltration decreases as the top soil gets saturated, and finally gets stabilized at a particular rate depending on particle size/texture of the soil.

The infiltration rate is a measure of how much water a soil can soak up in a given period of time. As should be expected, the more permeable a soil, the greater its infiltration rate. Hence, clay soils are characterized by low infiltration rates and sandy soils by high. The typical infiltration rates (mm h^{-1}) for various types of soils are as follows (Stern, 1979, p.88):

clay: 1 – 5; clay loam: 5 – 10; silt loam: 10 – 20; sand loam: 20 – 30; sand: 30 – 100.

The infiltration rate determines how long it will take for the rains or irrigation to water a unit area of land having a particular soil. For instance, a 60 mm precipitation or irrigation will take about 12 hours to water a unit area of clay soil, but only 2 hours to water the same areal extent if it is composed of sandy loam. In the case of clay soils which develop mud-cracks during the dry season, infiltration takes place rapidly to start with, but as the soil cracks get sealed, and the soil swells, infiltration slows down, before stabilizing. The infiltration characteristics of a soil are an important consideration in irrigation. When irrigation is applied in excess of the infiltration rate, water may be wasted (this is often the case where water cess is charged per hectare, but not on the basis of water used). Besides, the flowing water may cause erosion or may form puddles, from which it evaporates faster.

6.5.2 Irrigation Methods

Figure 6.7 shows the pattern of infiltration under alternative irrigation methods.

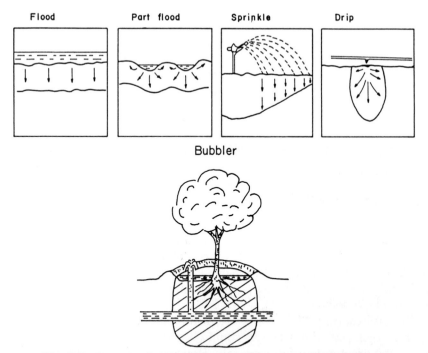

Fig. 6.7 Pattern of infiltration under different irrigation methods (*source:* Hillel, 1987, p. 61 © World Bank, Washington, D.C).

The principal methods of irrigation and their advantages and disadvantages are described below.

(i) *Surface irrigation or gravity irrigation*, whereby water introduced at the head of the field spreads and infiltrates throughout the field through forces of gravity and hydrostatic pressure. In this method, the soil is the medium through which water is *conveyed, distributed, and infiltrated*. For this reason, the physical and chemical properties of the soil, and the way the land has to be prepared for irrigation enter the picture prominently. This is the most ancient of all irrigation technologies and it has been estimated that this method serves about 95% of the irrigated land worldwide. This method is mechanically simple, has low energy requirements (as it is based on gravity), and can be easily adapted to small holdings. Its serious deficiencies are low application efficiency, high conveyance losses and wastage of water, which lead to adverse consequences such as waterlogging and salinization.

(ii) *Sprinkle irrigation*, whereby sprayed water falls on plants like rain. This system does not depend upon the soil surface for conveyance and distribution of water. It is hence not necessary to level the land (as required in the case of surface irrigation) and conveyance losses are minimal. Water can be applied at a rate less than that of the soil infiltrability. This allows soil aeration. The disadvantages are the high capital costs, maintenance requirements, and energy costs (for maintaining high pressure). Also, sprinkle irrigation is strongly affected by wind.

(iii) *Drip irrigation*, whereby water is applied directly to the root zone through a set of polyethylene tubes laid along the ground or buried at a depth of 15 to 30 cm. Water is delivered drop by drop through perforations or emitters in the pipes. The trickling rate is maintained at less than the rate of infiltrability in the soil. The operating pressures are 15 to 45 psi. Under this system, it is not necessary to level the land, and it is not affected by wind. By providing the rooting volume with the requisite amounts of water (and nutrients, which can be added to water), the soil can be kept continuously moist and aerated. As only a portion of the soil surface is wetted, the amount of direct evaporation is reduced. Slightly brackish water (up to 1000 mgL^{-1}) can be used under the method for some crops which are not too sensitive. As the water does not come into direct contact with foliage, it will not cause saline scorching. As the chosen volume of the soil is kept constantly wet, the salts in the brackish water do not have a chance to concentrate and affect the crop.

Drip irrigation can be used any where — in rugged terrains, in sandy soils of low moisture storage capacity, and arid climates of high evaporativity.

The capital costs of a drip irrigation system are high. The method needs very rigorous adherence to scheduling and maintenance. In an effort to cope with the extreme scarcity of water, Israelis have developed drip irrigation into a fine art. In the context of rising cost and scarcity of water, and lower costs of plastic tubing, this method is likely to be widely adopted.

A microprocessor-based, drip irrigation system suitable for use in developing countries, is now commercially available. This device uses low-cost ceramic sensors and operates on a solar cell-charged battery. The device continuously monitors the soil moisture and controls the drip rate so as to maintain the moisture within the desired limits.

(iv) *Microsprayer*: Microsprayers have several of the advantages of the drip systems in that the water is applied only to a fraction of the ground surface, high frequency irrigation is possible, and fertilizers can be injected, if desired. Besides, they have the following advantages over the drip systems: (i) as microsprayers have larger nozzle orifices, clogging of emitters is not a serious problem, and it is not necessary to filter the irrigation water; (ii) microsprayers are operated at pressures of the order of 2 atmospheres, which are much lower than for the drip systems; and (iii) microsprayers can be scaled down for use in small farms in the developing countries.

Microsprayers have the following minor disadvantages: (i) as wetting of leaves will be involved, brackish water cannot be used in irrigation; and (ii) since the area wetted is larger than in drip systems, there are some evaporational losses.

(v) *Low-head, bubbler irrigation*: A closed-conduit irrigation system has the great advantage in that it avoids conveyance losses and maintains uniformity in application. But any such system needs energy to pressurize water for distribution. Low-head, bubbler irrigation has the advantages of a piped system, but requires neither pumps nor nozzles. Even the low head available from a surface ditch may be adequate. In this arrangement, water is simply allowed to bubble out from open, vertical standpipes, 1 - 3 cm diam., which rise from buried lateral irrigation tubes. Bubbler systems are particularly suitable for widely spaced crops such as fruit trees and grapevines. Small circular basins can be constructed around the trees for the water to bubble into. Low-volume, high-frequency, partial area irrigation is possible with the system. The initial cost of the system (about USD 1600 ha^{-1}) is comparable or even lower than other systems. As the system works by gravity, there are no energy costs. It also has a longer life because it is buried. Rawlins (1977) gave a techno-economic analysis of the system.

6.6   IRRIGATION SCHEDULING

In the course of irrigation, water is introduced into that part of the soil profile that serves as the root zone of a crop plant. The soil moisture thus produced serves as a kind of bank providing water to the roots as needed. An ideal irrigation system is the one wherein the spatial and temporal distribution of water in the soil is just what the crop needs to achieve the highest productivity.

The timing and quantity of water application are decided on the basis of monitoring the soil, the plant, and the microclimate.

As the days pass after rains or irrigation, the moisture reserve in the root zone steadily reduces due to evaporation and root extraction. It is critically important to ensure that the soil moisture never goes down to the level of permanent wilting point, lest it adversely affect plant yield. Hence it would be prudent to keep the allowable depletion midway between the field capacity (PC) and the permanent wilting point (PWP). If the field capacity is 20%, and the wilting point 8%, the irrigator should apply water when the field capacity has dropped to 14%.

Monitoring soil needs information on the following aspects:

(i) Rooting depth of the crop: it may be shallow (0.3 – 0.7 m for vegetables), medium (0.9 – 1.5 m for wheat) or deep (1.0 – 2.0 m for maize, sorghum, and sugar-cane); (ii) potential and actual water contents of the root zone, and field capacity (expressed as fractional volume), on the basis of determination of soil moisture content, moisture tension, and moisture release characteristics of the soil (Fig. 6.8).

As soil moisture diminishes, the tensiometers record increased tension. This information can be used to estimate when the plants will suffer stress and need application of water. Because of the complexity of the variables involved, laboratory measurement of soil moisture may give misleading results. Soil moisture is best determined in *situ* with a portable neutron moisture meter. Though expensive, it has several positive features: measurement is made in *situ*, it can be made conveniently and instantly, and it gives meaningful results. Moisture tension is determined with a tensiometer (Hillel, 1987, p.37 & 38).

An experienced farmer or agronomist can assess the water status of the crop just by direct visual inspection of the foliage. Irrigation scheduling has to take into account the meteorologically imposed evapotranspirational demand. Time-domain reflectometry (TDR) is being increasingly used to determine irrigation scheduling (Topp and Davis, 1985).

6.7 IRRIGATION EFFICIENCY

The World Bank estimates that irrigation efficiency (i.e., net amount of water added to the root zone divided by the amount of water drawn from some source) is almost always below 50% (it may be as low as 30%). It is demonstrably possible to achieve an irrigation efficiency of 85 to 90%

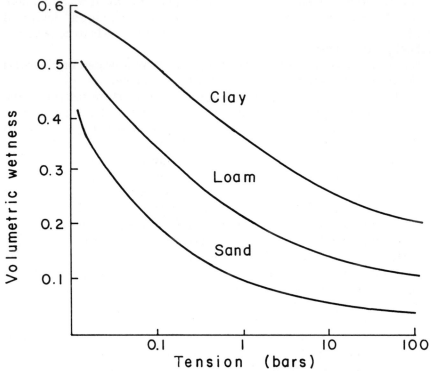

Fig. 6.8 Moisture release curves for various soils
(*source:* Hillel, 1987, p.35 © World Bank).

by proper management practices, particularly conveyance systems. It is therefore cost effective to spend money on improving irrigation efficiency rather than on constructing new irrigation reservoirs.

The efficiency of any process is usually taken as the measure of the output obtainable from a given input. In the case of irrigation, this could be defined in financial, physiological, sociological, etc. terms. Thus, irrigation efficiency may be considered the financial return for a particular amount of investment in water supply. This can vary tremendously from year to year and from place to place. Besides, it is not always possible to quantify the long-term sociological benefits of irrigation. In drought-prone situations, where the incidence of drought is not predictable, even a modest contribution to food security arising from assured irrigation can have very profound consequences on the quality of life of the people.

The conceptual basis for different kinds of irrigation efficiencies is shown schematically in Fig. 6.9.

The water supply efficiency is estimated in terms of flow at four points: (i) A-Flow at source, (ii) B-Flow at turnout, (iii) C-Flow into the field, and

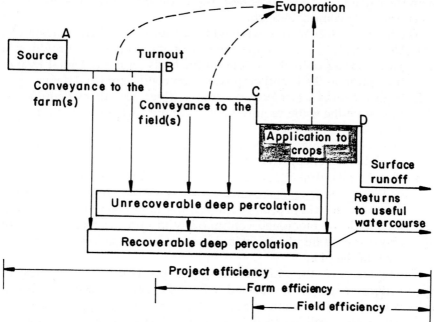

Fig. 6.9 Conceptual basis of different kinds of irrigation efficiencies (*source:* Chris Barrow, 1987, p. 108 © Longman's).

(iv) D-Return flow, i.e., excess, "used" or waste water from the scheme (some water may leave the system through subsurface flow).

Project efficiency deals with the relative proportion between flow at source and the amount of water applied to crops. Farm efficiency deals with the proportion of water at the turnout to the amount of water applied to crops. Field efficiency deals with the proportion of water flowing into the field to the amount of water applied to crops.

6.8 METHODOLOGY OF OPTIMAL IRRIGATION MANAGEMENT

Problems in the case of irrigation arise from a fallacy in human thinking — if something is good, more of it should be better. If a certain amount of irrigation raises crop productivity, which it surely does, more irrigation should produce more crops. As Hillel (1987) put it perceptively in his monograph, "The Efficient Use of Water for Irrigation", *Just Enough is Best — no less and certainly no more.*

For too long, irrigation has been thought of as simply a water delivery system, and this has had disastrous consequences. None would disagree with the dictum, *Just Enough is Best*. But to determine what is enough is an enormously complicated job.

The following scenario is envisaged for optimizing an irrigation system:

1. Determine the water requirements of the crop to be grown in a particular agroclimatic setting on the basis of the Potential Evapotranspiration (PET) and empirical crop coefficient (KC) data;
2. Withdrawal of water from the source (river, reservoir or aquifer) in tune with the water requirements, on time-variable and space-variable basis;
3. Delivery of water in tune with the rooting depth of the crop and available water capacity of the soil;
4. Planning the drainage system right at the outset as an integral part of the of the irrigation system. This is absolutely essential in the case of river valleys prone to high water-table conditions. Soil salinity should be monitored continuously in order to alert the farmer about the accumulation of injurious levels of salinity in the root zone of the plant.

The two facets of the traditional irrigation procedures are: (i) timing of application: irrigation is applied when the soil moisture is almost exhausted and (ii) quantum of application: irrigation is applied until the soil root zone is refilled to field capacity. The new methodology has different answers to the same questions: (i) timing of application: irrigate as frequently as possible, even daily, and (ii) quantum of application: irrigation is applied in quantities sufficient to meet the current evaporative demand and to prevent salinization of the root zone (Hillel, 1987, p.33).

In some countries, irrigation water is released on fixed dates in specified quantities. The farmer has to take it or leave it. Almost invariably, he takes it, just to be on the safe side. This often results in over-irrigation in some areas, with consequential problems of disposal of return flow, waterlogging, elevation of watertable, salinization, etc. The farmer has no incentive to conserve water. He pays water cess at a flat rate in proportion to the area of land. So a farmer near the top end of the irrigation scheme uses water excessively, whereas the tailenders complain that their lands do not receive enough irrigation. It is not uncommon for irrigation water to be highly subsidized.

There are two ways to ensure that irrigation water is not wasted: (i) irrigation water should be available on demand. For instance, nobody would buy postage stamps needed for six months, for the simple reason that nothing is gained by doing so; one can buy any quantity of stamps at any time without hassles; (ii) irrigation water should be properly priced in proportion to the quantity used per unit area of land. Pricing could have a built-in incentive to consume less water — *if a farmer uses less water per unit area of land, he should get the water at a cheaper rate.*

The recommended methodology of irrigation management involves the optimization of quantity and increasing the frequency of irrigation, with small daily rather than massive weekly or monthly application of irrigation water. The high frequency-low volume irrigation technique has profound consequences:

(i) Since the added water is small, it will only wet the few cm or few tens of cm of the surface. The flow below this point is essentially steady. The rate of through-flow through the soil, leaching of soil salts, and salinization consequences can be controlled. It becomes possible to maintain a wet root zone, while at the same time minimizing the drainage rate.

(ii) Infiltration, rather than extraction, would in consequence become the dominant process. Since water is supplied to the crop at nearly the precise rate needed by the plant, the ability of soil to store water and supply it to the plant when there is no irrigation or rain, becomes a property of minor importance. Thus, the "water-holding capacity" or "field capacity" of a soil need no longer be the constraining factor in irrigability of a given soil. Sandy or gravelly soils have very little moisture-holding capacity. Hence, even a short interruption in providing irrigation may result in deprivation, and severe stress. It is perfectly possible to irrigate coarse sands and gravels, even on slopes, by using drip or microsprayer irrigation.

6.8.1 Directions of Future Development

Irrigation units in developing countries have an enormous range — from several thousand hectares in the case of Government-sponsored commercial farms, to less than a hectare in the case of small family units. For instance, high-pressure sprinkle systems are cost effective in the case of large farms, whereas microsprayer or bubbler systems may turn out to be more suitable for family units.

It is not possible to design a universally applicable system of efficient use of water in irrigation because of the complexity of the variables with regard to soil, water, climate, crop and people, and because of the need to be compatible with other inputs such as seed varieties, fertilizers, tillage, pest control, etc.

Advances in information technology have made it possible to optimize the various system variables. Efficiency of water delivery needs to be optimized in terms of conveyance of water with minimal losses (say, in closed conduits), capability to provide measured amounts of water calibrated to meet the needs of crops in time and space, while preventing wastage, salinity, and rise in the watertable. Efficiency of water utilization is to be optimized to low-volume, low-pressure, high-frequency, partial-area irrigation to achieve high crop yields (Hillel, 1987, p. 99). Farmers in countries like the USA are already using high-tech systems whereby computers ensure delivery of the precise requirements of irrigation and nutrients to crop plants, as sensed by satellites in a pixel area (of about 25 m^2) in the context of a particular weather setting. Thanks to the availability of inexpensive PCs, groups of small farmers in the developing countries should also be able to avail of this approach, wherever the irrigation source is decentralized and compact, and directly under their control.

Irrigation is not a just biophysical-economic process. It is a human activity, carried out in a particular sociological setting. There is little doubt that modern irrigation is going to be knowledge intensive, and the principal input, which would also be the cheapest input, would be education of the farmer.

CHAPTER

7

Soils as Engineering Materials

Where there is no vision, the people perish

—Book of Job.

7.1 INTRODUCTION

A soil is defined differently in geology, pedology, agriculture, and geotechnology. For geotechnical purposes, soil is defined as: "Any naturally occurring deposit forming part of the earth's crust which consists of an assembly of discrete particles (usually mineral, sometimes with organic matter), that can be separated by gentle mechanical means, together with variable amounts of water and gas (usually air)" (Head, 1992, p. 360).

In a geotechnical sense, soils are engineering materials, but with one basic difference vis-á-vis other such materials. A concrete or a metal could be custom-made to suit a particular need. A soil, however, is a natural, low-unit-cost material. It therefore follows that a soil has to be used in its natural condition, with little or no processing, and as near as possible to the site of its occurrence.

There is wide variation in the physical, chemical, and geotechnical characteristics of soils. No two sites have identical soils. It is therefore necessary to study the soil conditions at a site, to determine how the soil is to be used.

Most of the soil tests are empirical, and are based on practical experience. In this chapter, two broad categories of tests are given — those which can be done by village artisans, where the soils are to be put to low-cost uses involving simple technologies (such as single-story housing with stabilized soil blocks in the rural areas), and more elaborate tests where soils are to be used for more sophisticated purposes involving stringent specifications.

7.2 CHEMICAL CHARACTERIZATION OF SOIL

Major elements in soils can be analyzed by standard wet chemical methods or by instrumental methods, such as AAS or XRF.

Table 7.1 gives the chemical analysis of some tropical (but not necessarily representative) soils from Jamaica, Kenya, Sudan and Egypt.

Table 7.1 Chemical composition of some tropical soils (*source:* ILO Technical Memorandum no. 12, 1987, p. 36)

Oxide	Jamaica red	Kenya red coffee	Sudan black cotton	Egypt
SiO_2	62.50	36.20	76.80	51.30
Al_2O_3	17.20	32.90	9.18	18.30
Fe_2O_3	8.39	10.72	3.54	8.19
Na_2O	1.13	0.27	0.33	3.32
K_2O	0.25	0.36	0.45	1.17
TiO_2	0.93	1.52	0.68	0.98
CaO	0.35	0.41	1.85	2.59
MgO	0.55	0.24	0.46	1.79
Mn_2O_3	0.04	0.33	0.05	0.05
SO_3	0.01	0.01	0.01	0.83
P_2O_5	0.01	0.22	0.01	0.15
LOI*	9.40	18.10	6.24	11.66

*LOI = Loss on ignition

The data are examined for purposes of determining their suitability for making stabilized soil blocks.

(i) The sum of fractions of silica, alumina, and iron oxide should be greater than 75%. This requirement is satisfied by all the four soils.

(ii) The percentage loss on ignition (LOI) should be less than 12%. High LOI indicates the presence of organic matter. For this reason, the Kenya soil with LOI of about 18% is not suitable.

(iii) Soluble salts in clay have a bearing on the plasticity, long-term strength, and corrosivity of soil blocks. Hence $Na_2O + K_2O$ should not exceed 2%. Thus, the soil from Egypt with $Na_2O + K_2O$ of about 4.5% is not suitable.

(iv) Soils are classified as lateritic and nonlateritic, depending on the ratio, $SiO_2/(Al_2O_3 + Fe_2O_3)$. A soil is called laterite if the above ratio is 1.33 or less, and termed lateritic soil if the ratio is in the range of 1.33 to 2.0. A non-lateritic soil has a ratio of 2.0 and above. Applying this criterion, the Kenya soil sample is a true laterite, the sample from Egypt is a lateritic soil, and the soil samples from Jamaica and Sudan are nonlateritic.

The other chemical tests and their relevance are summarized in Table 7.2.

Table 7.2 Procedures for chemical tests (*source:* Head, 1992, p. 236)

Type of test	Test procedure	Comments
pH value	Indicator papers Colorimetric (Kuhn) Lovibond comparator Electrometric	Simple test Available as kit Gives pH nearest to 0.2 Accurate to 0.1 pH
Sulfate content: Total sulfates; Water-soluble sulfates	Gravimetric method, Ion-exchange method	If the measured sulfate content is greater than 0.5%, the water-soluble sulfates should also be measured
Organic content	Dichromate oxidation	Measurement affected by presence of chlorides
Carbonate content	Gravimetric method; Collin's Calcimeter	Accuracy no better than 1% carbonate
Chloride content	Volhard's method; Mohr's method	Determination of water-soluble chloride content, employing silver nitrate
Loss on ignition (LOI)	Ignition	Destroys all organic matter

7.2.1 Significance of the Chemical Tests

pH value: Excessive acidity or alkalinity of soil water can adversely affect buried concrete. Even moderately acid soil can corrode the buried metal and other structures. Sometimes resinous materials are used for stabilization of soils for roads. While such materials may be used in the case of neutral or slightly acid soils, they are unsuitable for use in the case of alkaline soils.

Sulfate content: Sulfates in the soil can cause disintegration of concrete structures and corrode metal pipes which are in contact with the soil. The sulfates of sodium and magnesium, which are soluble, are more aggressive than the sulfate of calcium which is relatively insoluble in water. If the predominant sulfate present is that of calcium, determination of the total sulfate will lead to a misleading picture of the degree of corrosivity of sulfates in the soil. Hence, if the total sulfate exceeds 0.5%, the sulfate content of a 1:1 soil extract should be determined.

Depending on the sulfate content of the soil, the kind of cement that needs to be used for buried concrete to take care of the potential sulfate attack is given in Table 7.3.

Table 7.3   Cement type needed to take care of sulfate attack
(*source:* Head, 1992, p 28).

Total SO$_3$ (%)	SO$_3$ in 2:1 aqueous extract (g L^{-1})	Type of cement to be used
< 0.2	<1.0	OPC or RHPC, or with slag or pfa, PBFC
0.2 - 0.5	1.0 - 1.9	OPC or RHPC, or with slag or pfa; PBFC, SRPC
0.5 - 1.0	1.9 - 3.1	OPC or RHPC with slag or pfa, SRPC
1.0 - 2.0	3.1 - 5.6	SRPC
> 2.0	> 5.6	SRPC with protective coating

OPC = ordinary portland cement; RHPC = rapid hardening portland cement; PBFC = portland blast furnace cement; SRPC = sulphate resisting portland cement;   pfa = pulverised fuel ash.

Organic matter content: The organic matter content adversely affects the engineering properties of soils in the following ways: (i) reduction in bearing capacity; (ii) increase in compressibility; (iii) increase in possible swelling and shrinkage due to changes in moisture content; (iv) gases associated with organic matter can lead to significant settlement; (v) organic matter is often associated with acid pH and the presence of sulfates, both of which affect the foundations adversely; (vi) organic matter is undesirable where a soil is to be used for stabilization for roads.

Carbonate content: A high carbonate content means a low clay mineral content. Such a material has relatively high strength. The carbonate content is an index of cementing in compacted soils.

Chloride content: A high chloride content in coastal soils constitutes an indication that the soil has been affected by sea water (e.g. as in the Middle East). Aqueous solutions of chlorides are highly corrosive. Where such solutions are present, buried structures of iron and steel or steel-reinforced concrete have to be properly protected.

Whereas acid pH, presence of sulfates and chlorides in the soil invariably lead to corrosion of buried steel and concrete structures, corrosion is possible even in neutral or alkaline conditions when anaerobic, sulfate-reducing bacteria are present.

7.3 MOISTURE CONTENT AND INDEX TESTS

The moisture content of a soil has a critical bearing on the engineering properties of the soil. It has therefore to be determined in all cases. Other tests, called Index Tests, relate to the plasticity characteristics of the soil and are therefore performed in the case of cohesive soils.

All naturally occurring soils contain some amount of water. Such water is closely linked to platelike, fine-grained (< 2 μm) particles of clays.

The ways in which water exists in cohesive soils, and methods by which the extent of such association could be estimated, are summarized as follows (after Head, 1992, p.61):

(1) Water adsorbed on the surface of the particles: The water is firmly attached to the particle by powerful electrostatic forces and may be in the form of an extremely thin layer of the order of 50Å. It cannot be removed by oven drying at 110°C, and should be considered an integral part of the grain.

(2) Water that is less tightly held and hence can be removed by oven drying at 110°C. Water that can be removed by air drying is called hygroscopic moisture.

(3) Capillary water that is held by forces of surface tension and that can be removed by oven drying.

(4) Gravitational water that moves through pore spaces in the soil; this is the water involved in irrigation. It can be removed by drainage,

(5) Chemically combined water: Several minerals contain water of hydration within their crystal structure. Gypsum is one of the few minerals which loses its water of hydration in the process of oven drying. In most cases, oven drying cannot expel water of hydration.

Figure 7.1 shows the different ways in which water is associated with clay particles.

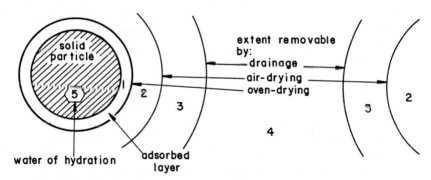

Fig. 7.1   Modes of association of water with clay particles in soil
(*source:* Head, 1992, p. 61 © Pentech Press, London).

For purposes of routine soil testing, moisture content is determined on the basis of loss of weight due to oven drying at 105 – 110°C:

$$w\ (\%) = (m_W/m_D) \times 100 \qquad (7.1)$$

where $w\ (\%)$ is the percent of moisture content, m_W is the mass of water removed by drying, and m_D is the mass of the dried soil.

7.3.1 Atterberg Limits

Atterberg Limits are named after the Swedish scientist, Atterberg, who first defined them in 1911 for agricultural soils. They comprise the liquid limit (LL), plastic limit (PL) and shrinkage limit (SL).

Every child who makes clay figures knows that a clay becomes plastic when moistened. With decreasing water content, the characteristics of clay undergo a profound transformation as follows (Head, 1992, p.62):

(i)   With excess of water, a clay slurry develops. The clay behaves like a viscous fluid. This is the "liquid" state.

(ii) With drying, the excess water is expelled. The clay particles are now able to stick together, and are in a position to offer some resistance to deformation. This is the "plastic state".

(iii) With further loss of moisture, the clay particles shrink, lose their plasticity, and become brittle. This is the "semisolid" state.

(iv) When no more drying is possible, and the clay shrinks to the minimum volume available, the clay is in the "solid" state.

Figure 7.2 shows the relationship between the water content, volume and shear strength, and the significance of various limits.

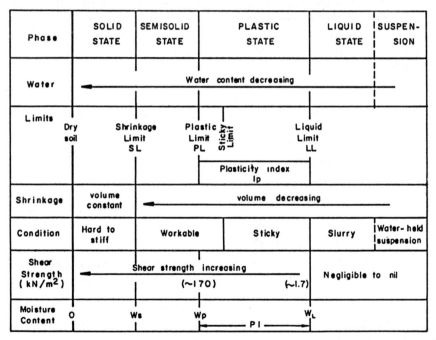

Fig. 7.2   Relationship between the water content, volume, shear strength and various limits
(*source:* Head, 1992, p. 63 © Pentech Press, London).

Liquid Limit (w_L) is the moisture content at which a soil passes from the plastic to liquid state. This is determined by the Cone Penetrometer Method, which is based on the measurement of penetration into the soil of a standardized cone of specified mass (80 ± 0.1 g). At the liquid limit, the cone penetration is 20 mm.

Plastic Limit (w_P) is the moisture content at which a soil passes from the plastic to solid state and becomes too dry to be in plastic condition. This test is relevant to soils which have some cohesion. It is determined on the minus 425 μm soil fraction. No special equipment is needed for this determination. About 20 g of prepared soil paste is well kneaded and shaped into a ball. It is rolled between the palms of the hands to form threads of about 6 mm diameter. The thread is placed on a flat, clean surface (glass, marble, metal) and rolled with the fingers until the diameter of the thread is reduced to 3 mm. If the thread can be reduced to 3 mm diameter, this indicates the presence of a high clay fraction. On the other hand, if the thread breaks up before being reduced to 3 mm, this indicates that the soil has a moderate sand fraction. The first crumbling point is the plastic limit. As soon as the crumbling stage is reached, the crumbled threads are gathered and the moisture content determined.

Plasticity Index (I_P) is the difference between the Liquid Limit (w_L) and the Plastic Limit (w_P).

Activity of clays: The particle size and mineral composition of a clay determine its Atterberg limits and the plasticity index. A given soil may contain clay (< 2 μm) mixed with coarser material. In such a case, the colloidal activity or simply, Activity, of a soil is obtained by dividing Plasticity Index (I_P) by the clay fraction (percentage of soil sample that passes a 425 μm sieve). The activities of different clays are as follows: Inactive clays (< 0.75), Normal clays (0.75 – 1.25), Active clays (1.25 – 2), Highly active clays (> 2). For instance, bentonite is highly active (6 or more).

Shrinkage Limit (w_S) is the moisture content at which a soil on being dried, ceases to shrink. Its measurement is based on the mercury displaced by a standard volume of the specimen (usually 38 mm diameter and 75 mm long) initially and after drying. The shrinkage ratio (R_s) is the ratio of the change in volume to the corresponding change in the moisture content above the shrinkage limit. The volume shrinkage of the soil (V_s, cm^3) is calculated from the decrease in moisture content from any value, w (%), to the shrinkage limit, w_s (%).

$$V_s = \frac{W - W_s}{R_s} \qquad (7.2)$$

Linear shrinkage (L_s) is the change in length of a bar sample of soil when dried from about its liquid limit, expressed as a percentage of the initial length.

About 150 g of soil passing 425 μm is mixed with distilled water and made into a paste. The paste is loaded into a greased, standard mold (usually with a length of 140 mm). The soil is allowed to dry slowly. The dried sample detaches itself from the mold. Then the soil is dried in an oven, first at 60 – 65 °C, and later at 105 – 110 °C. The length of the dried bar is measured.

The linear shrinkage (L_s) is calculated from the equation:

$$L_s\ (\%) = \frac{L_o - L_d}{L_d} \times 100 \tag{7.3}$$

where L_o is the original length (i.e., 140 mm, if the standard mold is used) and L_d is the length of the dry specimen.

Table 7.4 gives the typical ranges of the index properties of common clay minerals.

Table 7.4 Typical ranges of index properties of common clay minerals (after Head, 1992, p. 65)

Clay mineral	Liquid limit range	Plasticity Index range	Activity (approx.)
Kaolinite	40 - 60	10 - 25	0.4
Illite	80 - 120	50 - 70	0.9
Sodium montmorillonite	700	650	7
Other montmorillonites	300 - 650	200 - 550	1.5
Granular soils	20 or less	0	0

7.3.2 Empirical Tests

The physical and geotechnical properties of soils are based on the mineralogical composition, grainsize, chemical composition, and moisture content. The suitability of a given clay for use in earth dams, linings for ponds for storing irrigation and drinking water, canal bank linings, etc. can be evaluated on the basis of some simple, empirical tests which require no equipment and which can be performed even by artisans in villages. These tests lead to no numerical results but nevertheless are very useful. Head (1992, pp.109-114) gives a good description of these tests.

Puddle tests:

(i) Pinch test: The clay is kneaded well in the hands and formed into a ball of about 75 mm diameter. The ball is squeezed between the hands until it forms a disc of about 25 mm thickness. If there are no cracks in the disc, the clay has passed the test.

(ii) Tenacity test: The soil is rolled into a cylinder 300 mm long and 25 mm diameter. The cylinder is held up for 15 s vertically at one end so that the lower 200 mm of the cylinder is unsupported. If the clay supports its own weight, it has passed the test.

(iii) Elongation test: A soil cylinder, 300 mm long and 25 mm diam., is gripped horizontally at the two ends by the two hands, leaving 100 mm unsupported. The cylinder is then stretched until it breaks. The length of the neck formed at failure is noted. The longer the neck, the more suitable the clay. If the cylinder breaks with little or no stretching, the soil is not suitable.

(iv) Soaking test: The clay is made into a ball of 50 mm diameter. It is kept in a 600 ml beaker and covered with water. The state of the sample is observed at regular intervals of a few hours, up to 4 days. If the clay ball does not disintegrate, it is suitable.

Free swell test: This test was designed by Gibbs and Holtz. "Free swell" is the change in the volume of dry soil expressed as a percentage of the original volume. 10 ml volume of oven-dried, – 425 μm soil is slowly poured "in a drizzle" into 50 ml volume of distilled water. It may take anywhere from a few minutes to half an hour for the soil particles to settle down. The volume of settled solids increases (from the original volume of 10 ml) depending upon the swelling capacity of the clays. If the free swell value is more than 100%, this is an indication that the clay is an expanding type. Bentonites show very high free swell values, up to 2000%.

Sticky limit: This has been designed by Terzaghi and Peck. A clay is mixed with water such that it is plastic and "sticks" to a dry spatula blade. Then the clay is allowed to dry by exposure to atmosphere. The moisture content at the point when the clay is no longer sticky and the tool no longer picks up the clay, is the sticky limit.

7.4 PARTICLE SIZE DISTRIBUTION

A soil constitutes an assemblage of particles varying widely in shape and size, ranging from gravels to fine clays. Particle size analysis is done to determine the relative proportions, by dry mass, of different size ranges. The principal particle sizes and their corresponding particle diameters are as follows: Gravel (60 mm to 2 mm), Sand (2 mm to 0.06 mm), Silt (0.06 mm – 0.002 mm), Fines (< 63 μm), Clay (< 2 μm).

Sieving is used for coarse particles, i.e. of gravel and sand size. The ASTM sieve numbers and their aperture sizes are as follows:

no. 4 (4.75 mm), no. 10 (2.00 mm), no. 20 (850 μm), no. 40 (425 μm), no. 60 (250 μm), no. 140 (106 μm), no. 200 (75 μm).

Predominantly clayey and silty soils containing sand are first dispersed and then washed through a 63 μm sieve. The minus 63 μm fraction is collected and subjected to sedimentation test by pipette analysis or hydrometer measurement.

Particle size curves are drawn on the basis of the data obtained from the above analyses.

A triangular diagram can be used to depict the relative percentages of sand, silt and clay in a soil (Fig. 7.3). This diagram, however, will not be applicable for soils containing gravels.

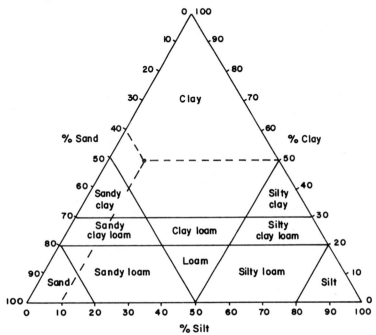

Fig. 7.3   Trilinear diagram of soil components
(*source:* ILO, 1987, p. 15)

The relationship between the particle size, mass, and surface area are given in Table 7.5.

The small cobble is the largest "soil" particle, and the clay is the smallest. The size range is enormous (75,000 to 1). With decreasing particle size, the mass of the particle decreases (by about 4.2×10^{14} to 1), and the approximate number of particles per gram and the approximate surface area increases.

Since soil particles are rarely spherical, their specific surfaces may be quite different from those given in the Table 7.5. Individual particles of clay generally tend to have platelike shapes, the surface areas being dependent upon the clay type. Thus the approximate surface area of

Table 7.5 Relationship between particle size, mass and surface area

Soil category	Particle size (nm)	Approx. mass of particle (g)	Approx. no. of particles g^{-1}	Approx. surface area (mm^2 g^{-1})
Small cobble	75	590	(1.7/kg)	30
Coarse sand	1	0.0014	720	2300
Fine sand	0.1	1.4×10^{-6}	7.2×10^5	23,000
Medium silt	0.01	1.4×10^{-9}	7.2×10^8	2.3×10^5
Clay	0.001	1.4×10^{-12}	7.2×10^{11}	2.3×10^6

kaolinite (10 m^2 g^{-1}) differs sharply from that of illite (100 m^2 g^{-1}) and montmorillonite (1000 m^2 g^{-1}).

7.4.1 Engineering Applications of Particle Sizes

Soils are used for several engineering purposes. Particle size analysis of soils has relevance in geotechnological applications in the following ways (Head, 1992, p. 169):

Selection of fill materials: Different zones of an earth dam or an embankment need filling material of different sizes.

Road subbase materials: In order to provide stable foundations, different layers of subbase of roads or airfield runways need to satisfy prescribed particle grade specifications.

Drainage filters: The grading specifications of drainage filters are defined in terms of the grade of the filtering layer above or below it or in relation to adjacent ground.

Groundwater drainage: This is largely determined by the relative proportion of fines (clay and silt fraction).

Grouting and chemical injection: Grading determines the most suitable grouting process, and the feasibility of grouting a particular ground.

Materials for making concrete, and for the fabrication of stabilized soil blocks: Grading determines the usability of a soil for these purposes.

Dynamic compaction: In some engineering situations, it may be necessary to improve the poor ground conditions by dynamic compaction. Grading would determine the feasibility of the process.

7.5 COMPACTION OF SOIL

Compaction of a soil, with or without stabilizer, increases its density, and improves other engineering properties (Table 7.6).

Table 7.6 Compaction of soil and its engineering benefits

Improvement in engineering properties	Beneficial effect on mass of fill
Higher shear strength	Greater stability
Lower compressibility	Less settlement under static load
Higher CBR (California Bearing Ratio) value	Less deformation under repeated loads
Lower permeability	Less tendency to absorb water
Lower frost susceptibility	Less likelihood of frost heave

Compaction curves (moisture content % vs. dry density, Mg m^{-3}) for five typical soils are given in Fig. 7.4.

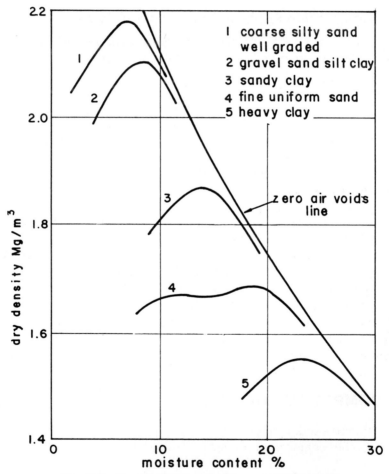

Fig. 7.4 Compaction curves for some typical soils
(*source:* Head, 1992, p. 314 © Pentech Press, London).

The Zero Air Voids line shown on the graph refers to the dry density-moisture content relationship for a soil containing no air voids.

It may be noted that the curves for most of the soils have clearly defined peaks, with the exception of fine sands which are characterized by a flatter compaction curve (sometimes this curve may have double peaks).

Engineering properties, such as shear strength, compressibility, etc. are measured for simulated samples which would have the dry density and moisture content *expected* of the compacted soil in the field. For some applications, more elaborate simulation tests have to be carried out to examine the consequences of changing conditions of applied stress.

Four types of compaction tests are made (after Head, 1992, pp. 319 – 339):

1. "Light" compaction test: In this method, the soil sample kept in a mold with a standard capacity of 1000 cm^3, is compacted by applying 27 blows with a rammer (2.5 kg) dropped from a controlled height of 300 mm. Compaction is calculated on the basis of the dry density and moisture content of the compacted sample.

2. "Heavy" compaction test: The procedure is the same as in the previous case, with the difference that the rammer is of 4.5 kg and the drop is 450 mm.

3. Compaction of stony soils: In the case of soils which contain gravel-size fragments larger than 20 mm, the CBR (California Bearing Ratio) mold is used. The mold has a nominal volume of 2,305 cm^3 and 62 blows are required.

4. Compaction using a vibrating hammer: In this test, granular soils passing a 37.5 mm sieve are compacted by a vibrating hammer (instead of a drop-weight hammer, as in the previous tests). The CBR mold has the dimensions of 152 mm diam. and 127 mm height, and the electric vibrating hammer (600 – 700 W) operates in the frequency range of 25 – 45 Hz.

7.5.1 Optimum Moisture Content

Let us examine the consequences of compaction under conditions of low and high moisture contents of a soil. If the moisture content of a soil is low, individual soil particles cannot come close to one another and there will be air voids. On the other hand, if the moisture content is high, there will no doubt be greater flow of particles when pressure is applied, but these particles will be separated by a film of moisture. When the soil dries, the films of water disappear, leaving behind air spaces. Thus, a too low moisture content and a too high moisture content are undesirable, because they lead to poor compaction and hence low density. Somewhere between the two extremes is the Optimum Moisture Content (OMC) which is the moisture content of a soil at which a specified amount of compaction will produce the maximum density (vide Fig. 7.5). The OMC and the density of a soil depend upon the type and quantity of stabilizing agent and the method and extent of compaction employed.

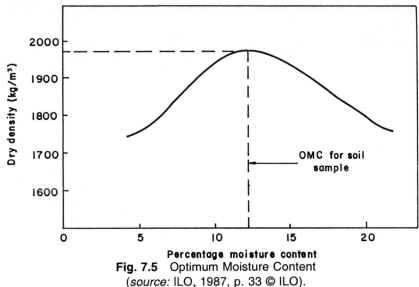

Fig. 7.5   Optimum Moisture Content
(*source:* ILO, 1987, p. 33 © ILO).

7.6   USE OF SOIL IN CIVIL CONSTRUCTIONS

Soil has been used to construct human shelters from time immemorial. Soil may be built into a wall *in situ* by building layer by layer through piling lumps of moist soil one above another ("cob" construction), or soil is daubed onto a framework of sticks ("wattle and daub"), or soil is rammed into the space between a pre-erected framework ("pisé de terre"). Figure 7.6 shows the various earth construction methods. Table 7.7 shows the purpose of soil stabilizers and how they work.

There are a number of advantages in using soils in civil constructions, particularly in the developing countries: (i) availability in large quantities in most regions; (ii) low cost (mainly for excavation and transportation) or no cost (if available at building site); (iii) easy workability, without needing specialized equipment; (iv) can be used for foundations, walls, floors and roofs, with appropriate modifications; (v) fire resistance; (vi) high thermal capacity, and low thermal conductivity and porosity makes it a suitable material for most climatic regions; (vii) low energy requirements for processing and handling unstabilized soil (compared to the 100 times more energy needed to manufacture and process the same quantity of cement concrete); and (viii) environmentally more appropriate

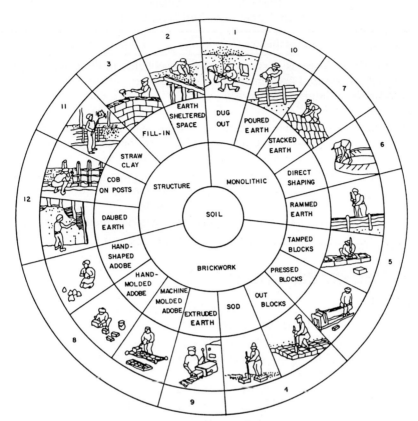

Fig. 7.6 Various earth construction methods
(*source:* Stulz & Mukherjee, 1993, © Swiss Centre for Development Cooperation in Technology and Management, Switzerland).

(because it uses an abundantly available resource in its natural form, with no pollution, negligible energy consumption, and no wastage) (Stulz and Mukherji, 1993).

Such structures, however, have a number of limitations: (i) in the case of the *in situ* method, cracks may develop in the finished structure due to shrinkage caused by the drying of moist soil. This not only reduces the strength of the wall, but also allows insects to lodge in the cracks; (ii) they do not survive for more than a few years in areas characterized by high humidity and rainfall; and (iii) they do not have a pleasing appearance.

Table 7.7   Soil stabilizers
(*source :* Stulz and Mukherji, 1993)

Purpose of the soil stabilizer	How it works
1. To increase the compressive strength and impact resistance of soil construction, and reduce the tendency of soil to swell and shrink	Stabilizer helps to bind the soil particles together
2. To reduce or completely exclude absorption of water (which could cause swelling, shrinking, etc.)	Stabilizer seals all the voids and pores, and covers the clay particles with a waterproofing film
3. To reduce cracking	Stabilizer imparts flexibility, by allowing the soil to expand or contract without cracking
4. To reduce excessive expansion and contraction	Fibrous material reinforces the soil

7.7   SOIL STABILIZATION

Stabilized soil blocks, prepared by the compaction pressure of moist soil mixed with a suitable stabilizer, are dense and strong, with well-shaped edges. As the stabilized soil blocks are allowed to dry before being used, walls built of such blocks do not develop cracks and thereby become weakened. Besides, because of their smooth surface, the blocks can be used without rendering or with a minimum use of rendering.

Use of stabilized soil blocks for the construction of walls, houses, subbase for roads, small bridges, earth dams, water tanks, etc. has much to recommend it, for the following reasons: (i) soil is locally available in most cases, either free or at a small cost; (ii) lime used in stabilization is often locally produced; (iii) the press can be manually operated. The fabrication process is therefore labor intensive and has a multiplier effect in generating employment downstream; and (iv) it does not require any costly inputs, such as expensive equipment, highly skilled labor, and investment and marketing strategies. Thus, manufacture of soil blocks is ideally suited for development as a microenterprise. In tune with the approach, tests to determine the suitability of a soil for being made into blocks are so chosen that they can be performed by village artisans (vide ILO Technical Memorandum no. 12, 1987, for details).

Some kinds of soil (such as alluvial soil, moorum, and clay) normally unsuitable for road construction, can be used for the purpose after being stabilized with lime. Slaked lime is spread over the soil surface and mixed thoroughly two or three times, so that a homogeneous soil-lime mix is formed. The layer is then thoroughly compacted with a power-roller, and cured for 7 days (source: *Rural Road Development in India*, v. II, 1990; Central Road Research Institute, New Delhi).

7.7.1 Soil Stabilization Methods

There are several methods of producing stabilized soil blocks, usable in the construction of walls, single-story houses, subbase for roads, earth dams, etc.

1. Manual or mechanical stabilization: When the soil has the Optimum Moisture Content (OMC), compaction will achieve the highest block density. To start with, the soil is moistened or dried as needed so that it will have the OMC level of moisture. The soil is then loaded into the mold and compacted manually or by machine. As can be seen from Fig. 7.5, when a compaction pressure of 3 MN m^{-2} is applied to a soil containing 50% silt and clay, and optimum moisture content of 12%, a density of 1,980 kg m^{-3} may be achieved.

Compacting pressures of 0.05 to about 4 MN m^{-2} can be achieved by manual methods, such as foot treading or hand tamping. In situations of small building loads, such as single-story houses, a compressive strength of 1.4 MN m^{-2} is considered sufficient. Thus, for rural housing and low-cost urban housing, stabilized soil blocks prepared manually may be satisfactory. In urban constructions with stringent specifications, higher compressive strengths may be needed. Thus, industrial production of soil blocks would require the use of mechanical equipment, which can achieve compacting pressures of several tens of MN m^{-2}.

2.Cement stabilization: Ordinary Portland Cement (OPC) is the most widely used cement in the world. When water is added to a mixture of sand and cement, hydration occurs, and the sand particles get embedded in the hard, cementitious gel. This process works effectively if a soil to be stabilized has a high sand fraction. If the clay fraction in a soil is high — and this shows up in increased shrinkage — the cement content in the soil mix has to be increased. In 1977, VITA (Volunteers in Technical Assistance) standardized the cement to soil ratio that needs to be maintained for soils of different measured shrinkages (expressed in mm):

Measured shrinkage (mm)	Cement to soil ratio
under 15	1 : 18 parts (5.56% cement)
15-30	1 : 16 parts (6.25% cement)
30-45	1 : 14 parts (7.14% cement)
45-60	1 : 12 parts (8.33% cement)

For a given strength of soil block, it is possible to reduce the percentage of cement needed by increasing the pressure applied. For instance, for a soil of measured shrinkage of 25 mm, 6% cement needs to be added, if the press available (say, Cinva-Ram of Colombia) is capable of just 2 MN

m^{-2}. By using a press capable of 10 MN m^{-2}, (say, BREPAK of U.K.), the percentage of cement needed can be brought down to 4%.

3. Lime stabilization: If a soil has a high clay content (with measured shrinkage of more than 60 mm), 6-8% lime can be used for stabilization. Both quicklime and hydrated lime can be used. The following chemical reactions take place when lime is added to moistened soil and mixed: (i) cation exchange — this will reduce the affinity of clays for water, thus leading to a lower moisture movement in the soil-lime mix; (ii) flocculation and agglomeration reactions which will have the effect of enhancing the viscosity of the mix; (iii) lime reacts with carbon dioxide from the air, thereby hardening the mix; and (iv) pozzolanic reaction — a chemical reaction between clays and lime, leading to the production of hydrated calcium-silicate-aluminate compounds. The first two reactions take place as soon as lime is added to the soil. Other reactions take a long time, even years. Pozzolanic reaction is temperature dependent and therefore fairly fast in tropical climates.

According to VITA, when lime is used as the stabilizer (instead of cement), its content should be double that of cement. However, doubling is not necessary if high compacting pressures are applied. For instance, wet compressive strengths of 3.0 to 3.5 MN m^{-2} have been achieved for the black cotton soil of Sudan (with a high silt and clay content of 58%) by using a compacting pressure of 8 to 14 MN m^{-2}.

The choice of cement or lime as a stabilizer is a question of availability and cost. While cement is an industrial product, lime can be produced locally by traditional methods. Generally, lime is cheaper than cement. In rural areas, only lime may be available. Thus, stabilization with lime is the recommended option for low-cost rural housing.

4. Stabilization with bitumen emulsion: Addition of bitumen emulsion to sandy soils produces strong waterproof blocks. But these are not recommended for tropical areas because of high costs and the tendency of bitumen to soften in tropical heat.

5. Gypsum plaster: Gypsum plaster can be used as a stabilizer for soils with medium levels of clay content. As they have poor water resistance, such blocks are best used for internal walls or for finishing internal wall surfaces.

7.7.2 Procedures for the Fabrication of Soil Blocks

The materials needed for the production of soil blocks are the soil, stabilizer (cement or lime or others) and water. The stabilizer is available in a powdered form, ready for use. It is therefore the soil that needs preprocessing. The soil may be too wet or too dry or it may contain stones and lumps. The soil is usually screened with a wire mesh screen which is kept inclined at an angle of 45° to the ground. Dry components (soil and stabilizer) are mixed first in the prescribed proportions. Water is added in small quantities, and the mixing is continued until all the required amount of water is used up. Hand-mixing gives perfectly satisfactory results.

The size of a typical block is 290 × 140 × 90 mm, and its production needs 7.5 to 8 kg of material. The quantity of materials needed to produce 300 blocks per day is as follows:

Material	8% hydrated lime	5% Portland cement
Soil	1.9 tons	1.95 tons
Stabilizer	150 kg	95 kg
Water	300 L	300 L
Total (after mixing)	2,350 kg	2,345 kg

A single-story house with a plinth of 50 m^2 needs 3,000 blocks.

A manually-operated block-making press, capable of compacting pressure of 10 MN m^{-2}, and daily production of 300 blocks of dimensions 290 × 140 × 100 mm, costs about USD 1500 and employs three people. There are cheaper presses (about USD 500) capable of 2 MN m^{-2}, and more expensive power-operated machines (about USD 20,000) capable of higher production (2,000 blocks d^{-1}).

The blocks are cured for periods depending on the nature of the stabilizer used (3 weeks for cement-stabilized blocks, and at least four weeks for lime-stabilized blocks). After curing, the blocks are tested for their engineering properties (dry and wet compressive strengths, water spray test, abrasive wear test, etc.)

It is now well-established that if there is good quality control with respect to raw materials, fabrication procedures, curing, etc., the soil blocks are not only much cheaper, but compare favorably with other kinds of walling materials in terms of engineering properties (Table 7.8).

Table 7.8 Range of properties of stabilized soil blocks and alternative walling materials
(*source:* ILO Technical Memorandum, 1987, no. 12, p. 5)

Property	Stabilized soil blocks	Fired clay bricks	Light-weight concrete blocks
Wet compressive strength(MN m^2)	1 - 40	5 - 60	2 - 20
Reversible moisture movement (per cent linear)	0.02 - 0.2	0 - 0.02	0.04 - 0.08
Density (g cm^3)	1.5 - 1.9	1.4 - 2.4	0.6 - 1.6
Thermal conductivity (W/m °C)	0.5 - 0.7	0.7 - 1.3	0.15 - 0.7
Durability under severe natural exposure	Good to very poor	Excellent to very poor	Good to poor

7.7.3 Natural Stabilizers

The most common *natural stabilizers* used in the developing countries, and their purposes are summarized below:

Straw and plant fibers: they check cracking in soils with high clay content. Soil can be reinforced by almost any type of straw (wheat, rye, barley, etc.) and also the chaff of most cereals. Other fibrous plant materials usable are: sisal, hemp, elephant grass, coir (coconut fiber), bagasse (sugarcane waste), etc. To achieve satisfactory results, the minimum proportion of plant reinforcement is 4% by volume — 20 to 30 kg per m^3 of soil is common. The straw and fibers should be chopped to lengths of not more than 6 cm and mixed thoroughly with the soil to avoid nests.

Plant juices: The juices of banana leaves precipitated with lime improves erosion resistance and slows water absorption. Latex of certain trees (e.g. *euphorbia, havea*) or concentrated sisal juice in the form of organic glue, reduces permeability of the soil.

Wood ashes: Ashes from fully burnt hardwood improve the dry compressive strength of the soil.

Animal excreta: These are mainly used to stabilize renderings. Cow dung reinforces the soil with fibrous particles.

Other animal products (horse urine, bull's blood, hair, glues, termite hills): these are not much used because of their low social acceptance.

ILO (1987) gives a detailed account of the whole process of fabrication of stabilized soil blocks.

7.8 GEOSYNTHETIC LINERS

7.8.1 What are Geosynthetics?

Geosynthetics are a group of fabricated products used for civil engineering purposes, such as reinforcement, as fllter or separation layer, or as a screen.

The principal polymers used in the manufacturing of geosynthetics are: Polyester (PET), Polypropylene (PP), Polyethylene (PE; there are two species: High Density-HDPE, and Low Density-LDPE), Polyamide (PA-polyamide products are not much used nowadays), and Polyvinylchloride (PVC).

The engineering characteristics of the polymers are given in Table 7.9.

Geosynthetics may be fabricated initially as membranes, tapes, threads, yarns, and fibers. These materials can then be woven, or stitched to non-woven or welded into various products: geotextiles (woven and non-woven fabrics together), geomembranes (two-dimensional sheets with very low permeability), grids, matting and composites.

Table 7.9 Engineering properties of polymers used in geosynthetics

Material	Unit mass, kgm^{-3}	Tensile strength at 20°C (N mm^{-2})	Modulus of elasticity (N mm^{-2})	Strain at break (in %)
PET	1300	800 -1200	12000-18000	8 - 15
PP	900	400 - 600	2000 - 5000	10 - 40
LDPE	920	80 - 250	200 - 1200	20 - 80
HDPE	950	350 - 600	600 - 6000	10 - 45
PA	1140	700 - 900	3000 - 4000	15 - 30
PVC	1250	20 - 50	10 - 100	50 - 150

The functional applications of different basic materials are given in Table 7.10.

Table 7. 10 Functional applications of the basic materials

Basic material	Reinforcement	Filter	Screeen
PET	x	x	x
PP	—	x	x
PE	—	x	x
PA	—	x	x
PVC	—	—	x

The engineering applications of the different geosynthetics are given in Table 7. 11.

Table 7. 11 Engineering applications of geosynthetics

Product	Properties	Applications
Woven fabric	Strength, stiffness, water-permeable, soil-retaining	Reinforcement, Filter
Nonwoven	Ductile/elastic soil-retaining, water-permeable	Filter
Geomembrane	Ductile/elastic soil-tight, water-impermeable	Screen

7.8.2 Geosynthetics in Civil engineering practices

Some important applications of geosynthetics in civil engineering are as follows (van Santvoort, 1995, p. 10).

Reinforcement: Geosynthetics are useful in building embankments and foundations for roads in situations where the natural ground has a low load-bearing capacity and in building structures on steep slopes.

Filter: Geosynthetics can be used to protect a bank with a filter construction and for separation of the embankment and natural ground.

Screen: Geomembranes are extensively used in sealing reservoirs, protection of roads in cuttings below the water table, and for sealing landfills.

Geosynthetics are generally environment friendly — they give off no emissions. However, care has to be taken to avoid pollution at the time of execution or during demolition of construction. It is necessary to ensure the flatness of the surface on which the membrane is to be spread. Geosynthetics may be affected by acid, alkali, oil, muck, and sunlight. Ultraviolet radiation and high temperature lead to aging of geosynthetics. Where geosynthetics are used to seal water reservoirs, care should be taken to ensure that they contain no toxic compounds which could leach into water.

7.8.3   Erosion Control by Biodegradable Coir Netting

From time immemorial, coir has been used in India to make ropes and matting. The Central Coir Research Institute (Government of India), Kalavoer, Kerala State, India 688 522, in cooperation with the Central Road Research Institute, New Delhi, India 110 020, developed innovative use of coir netting for erosion control. Plain woven netting or mesh of coir with square opening grids of 2.5 cm^2 is available in rolls of 1.3 m width. It can be used to protect such earthworks as dams, canal banks, highway and railway embankment slopes from erosion. The grids act like so many miniature dams. Experience has shown that if the coir netting is firmly laid on the sloping surface and grass seeds sown, the netting will hold the soil and seeds intact, thereby facilitating formation of a permanent vegetative turfing in due course. The technique has proven cost effective and successful in preventing surficial landslides in the Himalayan regions in the north, and Nilgiri Hills in the south of India, and protecting the highway slopes from erosion in parts of the province of Uttar Pradesh (northern India), etc.

Coir matting has a number of advantages:

 (i) It is inexpensive (coir mesh with 2.5 cm openings and 1 m wide costs about USD 0.60 cents m^{-2}).

 (ii) It is easy to install (to prevent displacement or possible undercutting by water, the top and bottom ends of the netting are anchored by pegging down the ends in 30 cm deep trenches).

(iii) It is biodegradable (when coir degrades after about 4-5 years, it adds rich organic matter to the soil at the rate of about 5 t ha^{-1}).

(iv) It is ideally suited to developing countries where coconuts are grown, because the coir mesh is made from locally available coconut waste, the operation is labor intensive and environment friendly.

7.9 SOILS IN WASTE DISPOSAL

The cardinal principle in the disposal of wastes, in particular hazardous wastes, is that they should be effectively isolated from the biosphere for a sufficient length of time, until they no longer present a risk to the biosphere. Two kinds of barriers are envisaged — the *natural* barrier of soil or rock which should prevent or keep within acceptable limits both the flow of water into the waste or the seepage of contaminated water from it, and the *technical* barrier composed of man-made material whose purpose is to seal the waste, in order to reduce leaching to a minimum, and to hinder chemical reactions of the waste material with the soil or rock which could adversely affect the capacity of the natural barrier to prevent contact with the biosphere (Archer et al., 1987).

It is necessary not only to contain the waste sites at their base and sides, but also to cap them to prevent the ingress of surface water.

The cheapest and the most common disposal site for domestic and commercial waste is an abandoned quarry/clay or shale pit/opencast coal mine/natural depression in the ground. It has to be ensured that the floor area of the pit is sufficiently large to allow a reasonable life for the site. An added advantage would be the availability of clay or similar kind of impervious material to seal the waste.

Where no suitable ready-made sites are available, a waste disposal site has to be excavated or a natural depression may have to be enlarged or deepened.

The two common methods of deposition of waste are either in the form of steeply dipping, uncompacted layers, or as thin, compacted layers covered by impermeable materials (Attewell, 1993, p. 93).

Toxic wastes need special kinds of sealing. Bentonite is the most commonly used natural sealant. It can absorb nearly five times its weight of water. When fully saturated, its volume expands 12-15 times its dry bulk. Na-bentonite is characterized by greater swelling than Ca-bentonite. The importance of bentonite to the waste disposal industry arises from the following considerations: (i) its high base-exchange capacity and large surface area (600-800 m^2 g^{-1}) enables it to capture and attenuate waste elements, especially heavy elements; (ii) because of its small particle size distribution, it can plug even the smallest voids against seepage; and (iii) its high liquid and plastic limits make it a very flexible structural component (Tewes, quoted by Attewell, 1993, p. 92).

To resolve the special containment problems of certain wastes, a new group of synthetic liners and sealants, collectively called *Geomembranes*, have been developed (see also section 7.7). These liners must have the following properties: (i) water-(and leachate-) tightness, (ii) toughness resistance to handling and puncturing, (iii) resistance to heat, ultraviolet light, and stress corrosion, (iv) resistance to penetration by roots and fauna, (v) resistance to permeation by hydrocarbons, especially chlorinated hydrocarbons, (vi) easy weldability on site, etc. The liners are usually made of polymer resins, such as High Density Polyethylene (HDPE). Some specialized geotextiles contain steel wires for protecting the linings against rodent attack. Attewell (1993, p. 114) describes how composite lining systems are to be organized for containment of the bottom and sides of a waste disposal site.

7.10 ENGINEERING APPLICATIONS OF LIME SLUDGE — A CASE STUDY

Limestone is a sedimentary rock, composed essentially of the mineral, calcite ($CaCO_3$). It constitutes about 15% of the sedimentary crust and hence is plentiful. It is produced in large quantities both in the industrialized and the developing countries. Lime is produced by calcining limestone at the temperature of 2200° F (1204°C). In this process, CO_2 is driven off, yielding lime:

$CaCO_3$ (limestone) + heat = CaO (lime) + CO_2 (carbon dioxide)

Lime is used extensively in the production of soda ash in the Solvay process, as a filler, as a flux in metallurgy, in glass and ceramic industries, in water treatment, etc.

Lime sludge is the waste material produced in the process of manufacturing lime. It constitutes roughly one-half of the quantity of lime. The sludge can be put to useful purposes such as agricultural liming, acid neutralization, flux for iron sinters, cement and glass industries, desulfurization of flue gases, etc.

Hart et al. (1993) gave a case history of the engineering applications of lime-sludge waste. A lime facility in northeastern Ohio produces about 220,000 t of lime and about 100,000 t of lime sludge per year.

The lime-sludge has the following average chemical composition (wt. percent): CaO – 51.48; SiO_2 – 8.30; Al_2O_3 – 3.24, Fe_2O_3 – 1.38; MgO – 0.94; Sulfur – 2.4%. Loss on ignition: 34.4%. Trace elements (mg kg^{-1}): Cr – 41, Pb – 10 and Se – 12.

Mineralogically, the sludge is a mixture of lime, calcite, gypsum, clay minerals, and pyrite.

The lime sludge has the following (mean) engineering properties: Natural water content: 90%, liquid limit: 48; plastic limit: 48; plasticity index: 0; Specific gravity: 2.57; maximum dry density: 64 pcf; optimum moisture content: 50%; permeability (cm s^{-1}): 2×10^{-5}; cohesion (psf): 1; friction angle: 27°; Unconfined compressive strength (psi): 38.

(62.4 pcf = 1 Mg m^{-3}; 0.145 psi = 1 kPa).

The high natural water content, low density and poor strength characteristics of the lime sludge render it unsuitable for engineering construction. In many engineering situations, natural soils do not have the required engineering properties and need to be stabilized by mixing with other soils or materials. The mixing of sludge with 30% soil increases the maximum dry density and decreases the optimum moisture content of the lime sludge to a level at which the mixture has the potential for use in engineering applications, such as daily landfill cover, hydraulic barriers, and structural fill materials.

The strength of the soft clay increases and its compressibility decreases when lime or cement is mixed with it. This is due to reaction of the clay with lime or cement, through processes of ion exchange and flocculation as well as pozzolanic reaction. The replacement of Na^+ ions by Ca^{2+} ions in the double layer surrounding each clay particle, reduces the size of the double layer and increases the attraction between the clay particles. This leads to a flocculated structure. A cementing gel forms when the silica and alumina in the clay mineral react with calcium silicates and calcium aluminate hydrates (pozzolanic reaction).

For clays with high organic content of more than 8%, the use of cement instead of lime, is recommended.

Addition of 5 to 10 % of quicklime has a profound beneficial effect on the geotechnical properties of soft clays — the unconfined compressive strength of the clays increases fivefold, and preconsolidation pressure threefold. The coefficient of consolidation increases by 10 to 40 times (Balasubramaniam et al., 1989, quoted by Miura et al., 1994).

Mud-built houses are a common sight in the villages in developing countries. The engineering characteristics of mud can be improved by mixing it with lime waste which is produced in the process of making hydraulic lime.

CHAPTER
8

Contamination of Soils

Every thing is connected to everything else, and every thing must go somewhere
> —Barry Commoner, in *The Closing Cycle.*

8.1   SOURCES OF CONTAMINATION

In this Chapter, the terms land, ground, and soil are used synonymously.

Soils get contaminated from various sources (Alloway and Ayres, 1993, p.18,19):

1. *Agricultural sources:*
Fertilizers — e.g. As, Cd, Mn, U, V, and Zn in some phosphatic fertilizers
Manures — e.g. As and Cu in pig and poultry manure
Pesticides — As, Cu, Mn, Pb, Zn, persistent organics (e.g. DDT, Lindane)
Corrosion of metals — e.g. galvanized metal objects (fencing, troughs, etc.)
Fuel spillages — hydrocarbons
Burial of dead livestock — pathogenic microorganisms

2. *Electricity generation:*
Ash, fallout — Si, SO_x, NO_x, heavy metals, coal dust.

3. *Derelict gaswork sites:*
Tars (containing HCs, phenols, benzene, xylene, napthalene, and PAHs), CN, spent Fe oxides, Cd, As, Pb, Cu, sulfides, sulfates.

4. *Metalliferous mining and smelting:*
Spoil and tailings heaps — wind erosion, weathering ore particles
Fluvially dispersed tailings — deposited on soil during flooding, river dredging, etc.
Transported ore separates — blown from conveyance, etc. on soil
Ore processing — cyanides, range of metals

Smelting — windblown dust, aerosols from smelter (range of metals).

5. *Metallurgical industries:*
Metals in wastes, solvents, acid residues, fallout of aerosols,etc. from casting and other pyrometallurgical processes.

6. *Chemical and electronic industries:*
Particulate fallout from chimneys
Sites of effluent and storage lagoons, loading, packing areas
Scrap and damaged electrical components, PAHs, metals, etc.

7. *General urban/industrial sources:*
Pb, Zn, V, Cu, Cd, PCBs, PAHs, dioxins, HCs, dumped cars, etc.
asbestos

8. *Waste disposal:*
Sewage sludge: NH_4^+, PAGHs, PCBs, metals (Cd, Cr, Cu, Hg, Mn, Mo, Ni, Pb, V, Zn, etc.)
Scrap heaps — Cd, Cr, Cu, Ni, Pb, Zn, Mn, V, W, PAHs, PCBs
Bonfires, coal ash, etc. — Cu, Pb, PAHs, B, As
Fallout from waste incinerators — Cd, PCDFs, PCBs, PAHs
Fly tipping of industrial wastes (wide range of substances)
Landfill leachates — NO_3^-, NH_4^+, Cd, PCBs, microorganisms

9. *Transport:*
Particulates (Pb, Br, Cl, PAHs), acid deposits, deicers, wide range of soluble/insoluble compounds at docks and marshaling yards and sidings, deposition of fuel combustion products, smoke, PAHs, SO_x, NO_x, rubber tire particles (containing Zn, Cd).

10. *Incidental sources:*
Preserved wood (e.g. PCP, creosote, etc.), discarded batteries (Hg, Cd, Ni, Zn), fishing and shooting (Pb), galvanized roofs, and fences (Zn, Cd); Warfare (e.g. fuels, explosives, ammunition, bullets, electrical components, poison gases, combustion products — PAHs),
Corrosion of metal objects — Cu, Zn, Cd, Pb,
Industrial accidents, e.g. Bhopal, Seveso, Chernobyl (wide range of pollutants).

11. *Deposition of atmospherically transported pollutants:*
Windblown soil particles with adsorbed pesticides and pollutants

(PAHs-Polycyclic aromatic hydrocarbons; HCs-Hydrocarbons
PCDD-polychlorodibenzodioxins; PCDF-Polychlorodibenzofurans).

8.2 ACID RAIN AND SOIL

The term, *Acid Rain*, was first used by Robert Angus Smith of England in 1872. The precipitation is naturally acidic because of the presence of carbon dioxide in the atmosphere, and it becomes more so because of natural emissions of sulfur dioxide and other species. The pH of the precipitation in remote areas is typically about 4.9, with a range of 4.0 to 6.0. In other words, even if there were no anthropogenic emissions, the pH of the precipitation would have been around 5.0. So strictly speaking, acid rain should refer to precipitation with pH of less than 5.

8.2.1 Sulfur Dioxide

Terrestrial and marine sources contribute to SO_2 emissions in roughly equal quantities. Anthropogenic emissions of SO_2 arise largely from the combustion of fossil fuels (coal, oil, and natural gas). Global natural sulfur emissions have been variously estimated at $50 - 100$ Tg S y^{-1}. They are of the same order of magnitude as anthropogenic emissions (about 80 Tg S y^{-1}) (one Tg = Tera g = 10^{12} g). Biogenic sulfur emissions are greatest in the tropical areas, whereas about 60% of anthropogenic emissions arise from North America and Europe.

Two aspects of sulfur dioxide emissions need to be recognized: emissions in a region are not uniform and may be highly localized. This is an important consideration while modeling transboundary pollution. For instance, over 75% of the emissions in the USA arise from the region east of the Mississippi River. Secondly, the regulations regarding atmospheric pollution being enforced in western Europe and North America, have yielded results in the form of reduced emissions.

8.2.2 Oxides of Nitrogen

The term NO_x comprises nitric oxide (NO) and nitrogen dioxide (NO_2). Apart from contributing to acidity, NO_x plays an important role in the photochemical reactions involving primary pollutants. Though both SO_2 and NO_x arise from the combustion of fossil fuels, there is an important difference between the two. Sulfur dioxide emissions arise from the sulfur contained in the fuel. NO_x arises not only from the nitrogen contained in the fuel, but also from oxidation of nitrogen in the air. Thus, NO_x emissions are influenced by the conditions during the combustion process, besides the composition of the fuel. Also, motor vehicle emissions contribute more NO_x relative to SO_2. The oxides of nitrogen are also produced by the burning of biomass (e.g. in the process of clearing the land for agriculture), with minor contributions from lightning, microbial activity, ammonia oxidation, etc. The total global emissions have been variously estimated from 25 to 99 Tg N y^{-1}. Natural sources account for about one-third of it, with anthropogenic sources contributing two-thirds. The spatial emission patterns of NO_x are broadly similar to those of SO_2, with Europe and North America contributing the major share of emissions.

8.2.3 Other Emissions

Though SO_2 and NO_x are the principal players with regard to acid rain, there are other minor sources contributing to acid deposition. Ammonia (NH_3) emissions from livestock wastes, fertilizer applications, and industrial processes are increasing in Europe because of intensive animal husbandry. They have a spatial pattern different from those of SO_2 and NO_x which are related to fossil fuel combustion. Some volatile organic compounds (VOCs) contribute to acid rain through their involvement in photochemical processes.

8.2.4 Deposition of Acidic Species from the Atmosphere

The atmospheric gases and aerosols have residence times ranging from a few minutes to a few days, and may be transported for distances ranging from a few km to a thousand km. Thus, a pollutant particle may get deposited at a place far away from its point of origin, depending on the nature and size of the aerosol and meteorological conditions.

The fallout of the larger particles follows Stoke's law. The velocity of the particles range from 9×10^5 to 1.3×10^{-1} cm s^{-1} for particles of diameter 0.1 to 20 μm.

The removal and deposition of the acidic species from the atmosphere may take place in three ways: (i) *wet* deposition, through precipitation as rain, snow, sleet, etc., (ii) *dry* deposition, direct deposition on terrestrial surfaces, and (iii) *occult* deposition, involving the capture by the impaction of cloud or fog droplets onto vegetation or other surfaces. Though the general public associate acid rain with wet deposition, acid rain should be understood to include wet, dry, and occult depositions. The ratio of wet to dry deposition depends upon the precipitation, and the dry faction may range from 0.2 to 0.9 of total deposition.

8.2.5 Environmental Consequences of Acid Rain

In southern Sweden, during a relatively short period of 30 – 60 years, the acidity of the soil in the forest areas has increased fivefold, nutrient levels have been halved, and acidification has penetrated to depths of about a meter. Even 30 years ago, nobody could visualize that this kind of development was possible, but nonetheless it has actually taken place.

In the normal course of events, the hydrogen ions (H^+) in the rain water which are rendered acidic by CO_2 in the atmosphere, are neutralized by Ca^{2+} and Mg^{2+} present in the soil humus. The lime in the soil thus acts as a natural filter which is continuously topped up with a certain amount of buffering substances produced by the weathering of rocks. However, if the input of acid substances in the soil exceeds the rate of replenishment of the weathering products, the filter will fail as it is unable to neutralize the acid inputs. When this happens, the acidic water will release such toxic metals as cadmium, copper, and aluminum which would have otherwise remained adsorbed on the sesquioxides and humus in the soil. Thus the most serious consequence of acidic water is that it dissolves metals both from soils and water pipes.

It is possible to differentiate between the water which has been made acidic by naturally produced carbonic acid (a weak acid) and anthropogenic strong acids. If the hardness/alkalinity ratio is close to 1, such water has not undergone anthropogenic acidification. If the ratio is greater than 2, it is a clear evidence that anthropogenic acidification is involved.

8.2.6 Mitigation of Acid Rain

Degradation of soil, water, and biota by anthropogenic emissions, has reached alarming proportions in the industrialized countries of North America and Europe. Some of the measures that have been taken by these countries to mitigate formation of acid rain, are given below.

As already stated, sulfur emissions arise principally from power-, and heat-producing combustion plants using coal or fuel oil with high sulfur content. Strenuous efforts are being made to bring down the emission in combustion plants to 0.05 g of S per megajoule (MJ) of fuel input, by using low-sulfur (< 0.2% S) fuel oil, and low-sulfur coal (< 0.5% S), desulfurization of flue gases, coal gasification technology, etc.

The bulk (about 70%) of the nitrogen oxide emissions is traceable to automobile emissions, with 20% coming from power and heat production, and 10% from industrial processes. As greater automobile emissions are associated with increasing speed and stop-and-go driving, reinforcement of speed limits is an obvious step for ameliorating the situation. Use of unleaded gasoline and fitting of catalytic converters in cars will also bring down nitrogen emissions from automobiles.

8.3 HEAVY METAL POLLUTION OF SOILS

There are two anthropogenic sources of contamination — primary sources, such as fertilization of soil in the process of working it, and secondary sources, whereby the pollutant gets added to the soil as a consequence of some activity in the neighborhood, such as smelting and aerosol deposition. In countries which have a long history of mining activity, contamination that arose from such activity persists even now. For instance, high levels of arsenic, mercury, cadmium, and lead in some soils in U. K. are traceable to historical mining. Nearly 85% of pesticide production in the world is used in the industrialized countries, but yet the pesticide-related problems in the Third World countries have been estimated to be 13 times more acute than in the industrialized countries. This is due to lack of adequate control in the production and application, storage, etc. of pesticides in the developing countries. Table 8.1 lists the various sources of soil contamination (source: Fergusson, 1990, p. 344).

Table 8.1 Sources of soil pollution

Source	Pollutant elements
Primary sources	
Fertilizers (e.g. phosphates)	Cd, Pb, As
Lime	As, Pb
Pesticides	Pb, As, Hg
Sewage sludge	Cd, Pb, As
Irrigation	Cd, Pb, Se
Manure	As, Se
Secondary sources	
Automobile aerosols	Pb
Smelters	Pb, Cd, Sb, As, Se, In, Hg
Refuse incinerators	Pb, Cd
Tire wear	Cd
Paint (weathered)	Pb, Cd
Marine	Se
Rubbish disposal	Pb, Cd, As
Long-range aerosols	Pb, As, Cd, Se
Coal combustion	As, Se, Sb, Pb
Chloroalkali cells	Hg

The extent of contamination of the soil from primary sources added to agricultural land is given in Table 8.2 (in terms of $\mu g\ g^{-1}$) (source: Fergusson, 1990, p. 344).

Table 8.2 Contamination of soil by primary sources $(\mu g\ g^{-1})$

Description	Pb	Cd	Hg	As
Phosphate fertilizer	4-1000	0.1-190	0.01-2	>1-1200
Nitrogen fertilizer	2-120	>0.05-0.1	0.3-3	2-120
Limestone	20-1250	>0.05-0.1		0.1-24
Sewage sludge*	2-7000	>1-56	>1-56	2-30
Manure *	0.4-16	>0.1-0.8	>0. 01-0. 2	>1-25
Irrigation water	<20	<0.05		<10
Pesticide (% element in pesticide)	11-26		0.6-6	3-30

* Dry weight basis

It can be seen from Table 8.2 that manure is the safest fertilizer, that the phosphate fertilizer pollutes more than nitrogen fertilizer, and that sewage sludge needs to be cleansed of its toxic components before being used for irrigation and fertilization. A high content of calcium in a phosphorite fertilizer buffers the consequences of the presence of toxic elements such as Cd.

Atmospheric deposition is more important for lead, arsenic and mercury, whereas cadmium is largely contributed by phosphate fertilizer. Cropping and drainage remove the elements from the soil. Some elements (such as lead, arsenic, and cadmium) build up in the soil, so that the input is greater than the output.

Quantification of the extent of pollution from secondary sources is rendered difficult because their contribution depends on the distance and intensity of the source and atmospheric conditions.

The behavior, reactions, and speciation of heavy metals in the soil are shown schematically in Fig. 8.1.

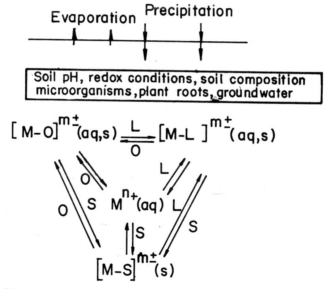

M - L = inorganic complex, M-O= organic complex
M-S = sorbed metal

Fig. 8.1 Behaviour, speciation and reaction of heavy metals in the soil (*source:* Brummer, 1986 © Spring-Verlag).

The mode of occurrence of individual heavy elements in the soils is summarized as follows (*source:* Fergusson, 1990).

Lead: The concentration of lead in the world soils has a range of 1 to about 900 μg g^{-1}, with a mean of 29 μg g^{-1}. The normal concentrations are in the range of 10-20 ppm and a concentration level of more than 100 ppm indicates contamination. Soils near metal-working units, high traffic roads, etc. may reach concentration levels of about 1,000 μg g^{-1}. Lead is the least mobile among the heavy elements. The half-life of lead in soils has a range of 800 to 6,000 years. Exchangeable lead is invariably below 5%. Soils contaminated as long ago as 4th century B.C., still show high levels of lead in the soil. Retention of lead is attributed to binding of the metal to organic matter. The principal ore of lead, galena (PbS), gets oxidized to $PbSO_4$ during weathering. The Pb^{2+} ion liberated in the process may get sorbed onto clays, organic matter and Fe-Mn oxides, or may be precipitated as insoluble compounds, or may form complexes with organic or inorganic ligands. Pb^{2+} may replace K^+ ions both in organic matter and clays. The order of sorption is: montmorillonite < humic materials < kaolinite < allophane < imogalite < halloysite, iron oxides. The sorption increases with increasing pH, up to the point when Pb $(OH)_2$ precipitates. Lead in the soil may be biomethylated, producing volatile and toxic $(CH_3)_4$ Pb and $(CH_3)_{4-n}$ Pb^{n+1}. The presence of lead reduces the enzymatic activity of the biota and, in consequence, incompletely decomposed organic material accumulates in the soil (Kabata-Pendias and Pendias, 1992). The speciation of lead depends upon the soil pH, source of lead (natural or anthropogenic), anionic species present, and the redox conditions. For instance, the lead compounds in automobile emissions, PbBrCl, PbBrOH, $(PbO)_2$, $PbBr_2$, get rapidly converted in the soil into insoluble compounds, such as PbO, Pb $(OH)_2$, $PbCO_3$, etc.

Cadmium: The content of cadmium in the soils has a range of 0.1 to 1.0 μg g^{-1}, with a mean of 0.62 μg g^{-1}. Soils with high organic matter content, such as Histosols, tend to have high Cd levels. pH is by far the most important factor affecting the speciation and mobility of cadmium. In acid soils, the solubility and mobility of Cd is determined by the organic matter, and Al, Fe, and Mn hydrous oxides in the soil. In high pH soils, such as calcareous soils or limed soils, Cd gets precipitated as $CdCO_3$. In soils high in chloride, cadmium could occur as cadmium chlorospecies, which tends to enhance the mobility of cadmium. Sorption of Cd is influenced by pH, ionic strength, competing ions, and constituents of the soil. The order of sorption is as follows: Al, Fe hydroxides, halloysite > imogalite, allophane > kaolinite, humic acid > montmorillonite >soil clay. The exchangeable metal is of the order 20 to 40%.

Increased accumulation of cadmium in soils (and consequent enhancement of its uptake by plants) is happening in several parts of the world. It has been estimated that when the cadmium content in the soil reaches a level of 3 ppm Cd, the food grown in such a soil will become unfit for human consumption. For instance, the soils in Switzerland receive input from the following sources: (i) atmospheric inputs: the atmospheric emissions of cadmium due to the burning of fossil fuels, smelting of ores, incineration of wastes, etc. have increased a hundredfold, with the consequence that the cadmium level in the upper 30 cm of soil has increased to 0.1 ppm (from the natural level of 0.01 ppm); (ii) direct discharges from sewage sludge and municipal waste compost; and (iii) phosphatic fertilizer. If the situation continues as it is now, the critical level of 3 ppm of Cd in the soil could be reached in 20 to 30 years (Fig. 8.2).

Mercury: The behavior of mercury in soils is influenced by (i) the volatility of elemental mercury, (ii) possible occurrence of free mercury in the soil, and (iii) biomethylation of mercury, leading to the production of toxic and highly volatile mercury compounds, such as $(CH_3)_2$ Hg. The concentration of mercury in most soils is of the order of $0.01 - 0.06$ μg g^{-1} (10 – 60 ppb). In some organic soils, the concentration could reach 400 ppb. Mercury contamination of soils may be caused by the mining of sulfide ores, production of dichlorine and caustic soda, and the use of mercury compounds as fungicides. The sorption of mercury would depend upon its chemical form, pH and Eh of the soil, other cations present, etc. Maximum absorption occurs wherever the soil contains large amounts of clays and/or organic matter. Among the clays, illite sorbs mercury better than kaolinite.

The background concentration of Hg in soils is about 70 ng g^{-1}. Near mining areas, mercury in soils may reach concentration levels of the order of μg g^{-1} (say, $\times$ 10 μg g^{-1}). Artisanal mining of gold invariably contaminates the soils seriously (as in Brazil and Tanzania).

Arsenic: The mean levels in relatively uncontaminated soils are 5-10 μg g^{-1}. Lowest levels are found in sandy soils derived from granites. Soils treated with arsenical pesticides may contain 600 μg g^{-1} (600 ppm) of arsenic. Under oxic conditions, arsenic occurs as arsenate $(AsO_4)^{3-}$. The species are strongly sorbed onto clays, Fe-Mn oxides/hydroxides, and organic matter. The amount of sorption is determined by the concentration of arsenic species, time and Fe-Mn content of the soil. Arsenite salts are 5-10 times more soluble than arsenate salts. Reducing conditions, such as those occurring in flooded soils (e.g. paddy soils), enhance the proportion of As (III), in which form As is more available and more toxic. Soil bacteria can accelerate the oxidation of arsenite to the less soluble arsenates. They can also bring about methylation, in the form of As (III) methyl derivatives.

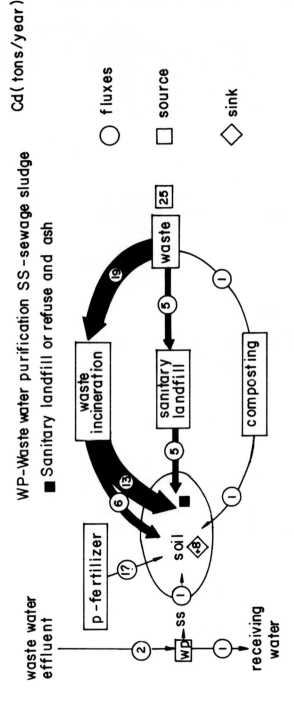

Fig. 8.2 Cadmium input into the soil from different sources, a case history in Switzerland
(*source:* Förstner, 1987 © Springer-Verlag).

8.4   PESTICIDE CONTAMINATION OF SOILS

Pesticides are widely used in agriculture, public health, and several other purposes. There are more than 10,000 formulations of pesticides, of which about 450 are widely used. The extensive use of pesticides during the last fifty years not only led to their becoming ubiquitous pollutants in soils, crops, groundwater, human and animal tissues, etc. in industrialized countries, but also led to their dispersal to far-off places; pesticide residues have been found in Antarctic penguins.

The main types of compounds used as pesticides are as follows (Manahan,1991):

Insecticides (insect killers):
Organochlorines (e.g. DDT, Lindane, Aldrin, Heptachlor)
Organophosphates (e.g. Parathion, Malathion)
Carbamates (e.g. Carbaryl, Carbofuran)

Herbicides (weed killers):
Phenoxyacetic acids (e.g. 2,4-D; 2,4,5-T; MCPA)
Toluidines (e.g. Trifluralin)
Triazines (e.g. Simazine, Atrazine)
Phenylureas (e.g. Fenuron, Isoproturon)
Bipyridyls (e.g. Diquat, Paraquat)
Glycines (e.g. Glyphosate "Tumbleweed")
Phenoxypropionates (e.g. "Mecoprop")
Translocated carbamates (e.g. Barban, Asulam)
Hydroxy nitriles (e.g. Ioxynil, Bromoxydynil)

Fungicides (fungus killers):
Nonsystemic fungicides
   Inorganic and heavy metal compounds (e.g. Bordeaux Mixture-Cu)
   Dithiocarbamates (e.g. Maneb, Zineb, Mancozeb)
   Pathalimides (e.g. Captan, Captafol, Dichofluanid)
Systemic fungicides
   Antibiotics (e.g. Cycloheximide, Blasticidin S, Kasugamycin)
   Benzimidazoles (e.g. Carbendazim, Benomyl, Thiabendazole)
   Pyrimidines (e.g. Ethirimol, Triforine)

As evidence mounted on the toxic effects of pesticides on human beings and animals, some pesticides, such as DDT, Aldrin and Dieldrin, were completely banned in the Industrialized countries. Though the ban was effected in the 1980s, these pesticides continue to persist in the environment to varying extents:

Pesticide	% decay	Time taken (years)
DDT	44	8
BHC	50	1
Dieldrin	49 – 53	3

In the place of the banned pesticides, substitutes such as pyrethroids and organophosphorus compounds, have been introduced in developed countries. These substitutes are environmentally more acceptable, though more expensive. For instance, pyrethrins (derived from the flower of the plant, *Chrysanthemum cinerariaefolium*); undergo rapid photolytically induced oxidation, and are therefore shortlived. Several synthetic analogues of pyrethrin have been developed. Intensive work is underway to develop a whole range of pesticides, based on the seeds of the Indian tree, *neem* (*Azirdirachta Indica*). Some of the organophosphorus pesticides, such as Chlorpyrifos, are highly effective against pests, but less toxic to mammals. They are applied in relatively small quantities (200 – 1,200 g ha^{-1}). They break down by hydrolysis and their persistence in the soil is in the range of 60 – 120 days.

Soil processes have great ability to decompose organic matter. The enzymatic activities of the microorganisms are capable of degrading and detoxifying a variety of substances added to the soil. However, the chemical structures of some pesticides are such as to preclude their enzymatic degradation. Several synthetic pesticides (such as 2,4,5-T or 2,4,5-trichlorophenoxyacetic acid) are heavily substituted with chlorine, bromine, fluorine or nitro or sulfonate groups which are not found in biological tissues. Microorganisms cannot degrade such pesticides, which therefore persist in the soil for long periods (say, about 15 years).

Those pesticides characterized by low vapor pressure and low solubility in water tend to persist in the soil. The extent of persistence depends upon the temperature, soil type, and soil microbiology. For instance, the herbicide, trifluralin, may persist in the soil between 15 to 30 weeks. Herbicides tend to be lost more readily from moist soil than from the same soil when it is dry. The molecules of herbicides that are strongly adsorbed on dry soils get dislodged under damp conditions.

Herbicides interact with soil components in complex ways. Urea derivatives, triazines, carbamates and nitrophenyl ethers, etc. tend to be strongly adsorbed on soil organic matter. Quaternary ammonium compounds are adsorbed on clays and metal hydroxides. The herbicides may also interact with the exudates from soil microorganisms.

The probable range of persistenc of some important herbicides is given in Table 8.3 (source: Hassal, 1987, p. 253). The term, probable, is used advisedly, because persistence is greatly influenced by the rate of application, rainfall, soil type, temperature and microbial population. Besides, some crops are more tolerant than others for the same herbicide.

Table 8.3 Time scale of persistence of some herbicides

Herbicide	Range of time scale of persistence (in weeks)	Usual dose of application (kg ha^{-1})
Propham	2.6-10	2.2
Dalapon	3-12	4.7
Aminotriazole	3-10	4.5
MCPA	6-14	1.4
Chlorpropham	6-20	1.1
Prometryne	8-16	1.1
Diallate	7-22	1.7
Triallate	10-25	1.7
Cycloate	13-24	2.0
Pyrazone	13-25	1.4
Monolinuron	13-37	1.1
TCA	14-40	1.7
Trifluralin	19-75	1.1
Bromacil	22-50	2.2
Atrazine	23-60	1.7
Chlorthiamid	24-75	9.2
Dichlobenil	25-80	9.2
Diuron	33-115	1.8
Simazine	33-125	1.1

The soil type not only influences the period of persistence of a herbicide, but also affects the toxicity of a given dose to the weeds. For instance, carbamate and urea herbicides are more toxic to the weeds in light soils than to the same weeds in heavy organic soils. Consequently, these herbicides persist in organic soils for longer periods.

8.5 FATE OF POLLUTANTS IN SOILS

Alloway and Ayres (1993) gave a good summary of this theme.

There is a basic difference between the behavior of pollutants in the soil vis-à-vis that in air and water. While the concentration of pollutants in moving air or stream water tend to decrease in due course through dilution or mixing, pollutants deposited on the soil tend to get concentrated due to adsorption processes which bind organic and inorganic pollutants to the surfaces of soil colloids.

The more tightly a pollutant is bound (to say, organic matter), the more it will get concentrated in the top layers of the soil, the less will be its ability to percolate to the water table and contaminate the groundwater. Also, a tightly bound pollutant becomes less bioavailable, which is a good thing. However, such an adsorbed pollutant cannot be easily decomposed, resulting in the persistence of the pollutant in the soil for longer periods.

On the other hand, in the case of sandy soils with low contents of organic matter and clays, and therefore of low adsorptive capacity, the pollutant can infiltrate down the soil and contaminate the groundwater.

The reactions taking place in the surface layer of sediments are similar to those occurring in the soil, as oxic conditions prevail in both cases. Gleyic (waterlogged) soils and deeper sediments are characterized by anoxic conditions.

The extent of adsorption of a pollutant depends on the composition of the soil (amounts and types of clay minerals, hydrous oxides and organic matter), pH-Eh conditions in the soil, and the properties of the pollutant. For instance, inorganic cations and anions, and some organic molecules (such as the bipyridyl herbicides) tend to get adsorbed on soil colloids. Humic polymers adsorb nonionic organic molecules, such as hydrocarbons, organic pollutants, and pesticides. If an organic pollutant is soluble, it will be readily leached in areas of good precipitation.

Organic micropollutants get degraded in two ways: (i) degradation through physicochemical processes such as hydrolysis, oxidation/reduction, volatilization, and photodecomposition; and (ii) microbial decomposition.

The extracellular enzymes associated with soil microorganisms are capable of decomposing most of the organic pollutants adsorbed on the soil, once the microorganisms have become adapted to a particular substratum. However, the microorganisms have difficulty in decomposing man-made organochlorine pollutants such as DDT, PCBs and PCDDs, because the carbon-chlorine bond contained in such pollutants is not found in nature. This explains the long persistence time (about 10 years) of such compounds. However, some species of bacteria and fungi have evolved the ability to decompose organochlorine compounds, and these are increasingly being utilized to ameliorate contaminated land.

There are cases wherein the decomposition products of some pollutants are more toxic to humans and animals than the original pollutant. For instance, the products of the microbial oxidation of benzo-[a]-pyrene, are carcinogenic because they bind to cellular DNA.

Several solvents, such as chloroform and 1,1,1-trichloroethylene can leach through the soil profile and contaminate the groundwater. Some chlorinated aliphatics do not get degraded, but get removed from the soil through volatilization.

Though insoluble organic pollutants adsorbed on soil colloids can rarely move down the soil profile (except occasionally through desiccation cracks or worm channels), they can be transported for long distances by winds. There has been one well-documented instance whereby soil particles transported from China caused pesticide pollution through brown snowfall in Arctic Canada. A pollutant adsorbed on a soil particle can enter the hydrological regime when it is carried piggyback in the runoff.

The maximum acceptable level of pesticides in potable water is 0.1 μg L^{-1}. Experience has shown that herbicides are the most important sources of contamination of drinking water.

8.5.1 Root Uptake of Toxic Metals from Soils

Toxic metals in soils find their way to plants and then to human beings. A simple, but not accurate way of estimating the concentration of a particular metal in plants is to multiply the total concentration of the metal in the soil by the chemical specific Root Uptake Factor (RUF) taken from the literature (Baes et al., 1984). This method does not take into account various soil variables, such as pH, organic matter content (OMC), and cation exchange capacity (CEC), which are known to influence the root uptake of metals. On the basis of multiple regression analysis of a large body of data related to studies made near the Globe Plant (a smelter) in Colorado, USA, Hattemer-Frey et al. (1994) developed predictive equations for estimating the root uptake of As, Cd, Cu, and Zn in home-grown fruiting vegetables (FVs, such as, cucumbers and tomatoes) and root vegetables (RVs, such as carrots) from the content of the metal in the soil, pH, OMC, CEC, etc. The coefficient of determination (R^2) is a good indicator of the extent to which the independent variables are related to the dependent variable.

Arsenic (As):
FVs = 0.007 soil + 0.22 pH; R^2 = 0.87

RVs = 10 + 0.22 soil + 1.3 pH – 0.9 OMC + 0.21CEC; R^2 = 0.99

This shows that the concentration of As in home-grown FVs increases with increasing pH (i.e., alkalinity) and increasing concentration of As in the soil. On the other hand, the concentration of As in RVs has a positive correlation with the content of As in the soil, pH and CEC, but a negative correlation with OMC.

Cadmium (Cd):
FVs = 0.019 soil + 0.11 pH; R^2 = 0.7

RVs = 0.54 soil – 0.48 pH + 0.29 CEC; R^2 = 0.93

The equations indicate that the Cd content in FVs is directly correlated to the concentration of the element in the soil, and the pH, whereas the concentration of Cd in RVs has a positive correlation with the concentration of the element in, and CEC of, the soil, and a negative correlation with pH (i.e., the Cd in the root vegetables increases with greater soil acidity).

Copper (Cu):

FVs = 0.08 soil + 1.62 pH − 1.037 OMC; R^2 = 0.99
RVs = 0.17 soil + 2.14 pH; R^2 = 0.97

The concentration of copper in FVs correlates positively with content of the element in the soil and pH, and negatively with OMC. The concentration of the element in RVs is positively correlated with the concentration of the element in the soil, and the soil pH.

Zinc (Zn):

FVs = 0.075 soil + 24.1 pH + 1.1 OMC − 0.01 Fe; R^2 = 0. 75
RVs = 0.21 soil + 5.9 pH + 15.15 OMC; R^2 = 0.91

The Zn content of both FVs and RVs is positively correlated with the concentration of the element in the soil, soil pH, and OMC, except that there is a negative correlation with Fe content of the soil in the case of FVs.

Thus, the concentration of As, Cu and Zn in the FVs increases with the concentration of the element in the soil and soil pH, except in the case of cadmium. It is known that Cd is more readily taken up by plants in an acid environment.

Table 8.4 gives the range of values and the mean for various soil parameters in the Globe Plant study (source: Hattemer-Frey et al., 1994). Though the values given in the Table pertain to the specific location and soil type, and are applicable to fruiting and root vegetables only, they do indicate the order of magnitude of metal values met with in soils and plants, and the factors controlling the pathways of metals in these media.

Table 8.4 Soil parameters in the Globe Plant (USA) Study

Soil parameter	Range and (mean)
As in soil (ppm, dry wt)	6.9 - 162 (27.8)
Cd in soil (ppm, dry wt)	0.5 - 112 (16.2)
Cu in soil (ppm, dry wt)	20 -183 (59.4)
Zn in soil (ppm, dry wt)	89 - 2325 (358)
Soil pH	6.5 - 8.9 (7.4)
CEC (m eq./100 g)	6.8 - 27.4 (14.1)
OMC (%)	0.8 - 4.8 (2.7)
Soil Fe (ppm, dry wt)	11,250 - 21,500 (15, 805)

8.5.2 Pathways of Radioactive Pollutants in Soils

The artificial radioactivity of soils arises from the sedimentation of nuclides. The principal nuclides involved are: ^{137}Cs, ^{134}Cs, ^{131}I and ^{90}Sr. The mobility of nuclides in the soil profile depends on the pH, Eh, texture (content of clay), porosity and permeability and organic matter content. The rate of movement ranges from a few mm to several cm per year. In the case of agricultural soils, the kind of tillage employed is of critical importance.

In the case of land plants, some radionuclides are picked up by the plants from the precipitation via the stomata (e.g. ^{95}Zr, ^{103}Ru, ^{106}Ru, ^{140}Ba, ^{140}La, ^{144}Ce, ^{144}Pr), as well as from the soil via their roots. The physiology of the plant and the geochemical environment of the soil would determine which plant would take up which nuclides and in what proportion. For instance, ^{91}Y, ^{95}Zr, ^{106}Ru, ^{127}Te, ^{141}Ce, and ^{239}Pu (which are mostly hydrolysate elements) tend to accumulate in the roots, whereas ^{90}Sr and ^{137}Cs end up in the aboveground parts. Thus, the kind of radionuclides that we should look for in (say) potatoes, may be different from those that we should look for in (say) tomatoes.

The movement of ^{137}Cs and ^{90}Sr from soil to plant is a case in point. The soil-plant transfer factors for Cs and Sr are given in Table 8.5 (source: Schmidt et al., 1986, quoted by Becker-Heidmann and Scharpenseel, 1990).

Table 8.5 Soil -plant transfer factors for Cs and Sr *

	Cs	Sr
Leafy vegetables	0.075 - 0.9	0.08 - 7.8
Potatoes	0.023 - 0.16	0.015 - 0.38
Root vegetables	0.0025 - 0.15	0.55 - 21
Grass	0.0011 - 14	0.018 - 9.8
Clover	0.004 - 33	0.22 - 7.4
Vegetation	0.05	0.4

* Ratio of Bq kg^{-1} of fresh plant material to Bq kg^{-1} of dry soil

On April 26, 1986 (Saturday), there was an explosion in one of the four graphite-moderated, water-cooled reactors at Chernobyl, 128 km from Kiev in the Ukraine (erstwhile Soviet Union). Chernobyl is on the banks of the Pripyat River which flows into the Dnieper. The region is relatively flat with gentle slopes. It is the most serious nuclear accident to date. About 50 MCi (million curies) or 2×10^{18} Bq of radioactive fission products and noble gases (including ^{123}Xenon) escaped. About half of this amount relates to nuclides which figure in the food chain. About 50% of the emissions of the condensable products fell in an area of about 60 km radius around the accident site (see Aswathanarayana, 1995, pp. 162-165).

Korobova et al. (1998) studied the mobility of ^{137}Cs and ^{90}Sr in soils, and their transfer in soil-plant systems in the Novozybbkov District affected by the Chernobyl accident. They found a sharp contrast in the behavior of ^{137}Cs which is strongly fixed in the soil (to the extent of 40 – 93%), whereas 70 – 90% of ^{90}Sr is present in water-soluble, exchangeable, and weak-acid-soluble forms. Vertical migration of radionuclides, which is detected to a depth of 30 – 40 cm, is most pronounced in local depressions with organic and gley soils. In woodlands, most of the ^{137}Cs is fixed in the plant litter and upper mineral soil layer. In the case of floodplain grasslands, radionuclides are associated with soils with fine texture. The uptake of radionuclides by plants decreases in the following order: legumes > herbs > grasses. A high accumulation of ^{137}Cs in potato tubers grown in sandy, podzolic soils, has been noted. The authors recommend that people living in the areas within the 'zone of contamination exceeding 15 Ci/km^2, should avoid eating local forest products, and cattle should not be allowed to graze on wet floodplain meadows.

8.6 ENVIRONMENTAL QUALITY CRITERIA FOR SOILS

The normal ranges and toxic levels (μg g^{-1}) of various heavy elements in soils are given in Table 8.6 (after Alloway, 1990).

Table 8.6 Normal ranges and toxic levels (μg g^{-1}) of heavy elements in soils

Element	Normal range	Toxic level
Ag	0.01 - 0.8	1 - 4
As (III)	0.02 - 7	5 - 20
Cd	0.1 - 2.4	5 - 30
Cu	5 - 20	20 - 100
Cr	0.03 - 14	5 - 30
Hg	0.005 - 0.17	1- 3
Ni	0.02 - 5	10 - 100
Pb	5 - 10	30 - 300
Sb	0.0001 - 2	1 - 2
V	0.001 - 1.5	5 - 10
Zn	1 - 400	100 - 400

Environmental quality criteria for soil (μg g^{-1}) for various purposes are given in Table 8.7 (source: Canadian Council of Ministers for the Environment, Winnipeg, 1991):

Heavy metals in soils are usually determined by solvent extraction or acid digestion, followed by estimation by atomic absorption spectrometry. Flame AAS is suitable in the concentration range of 0.1 – 10 μg mL^{-1}. At

Table 8.7 Environmental quality criteria ($\mu g\ g^{-1}$) for soils for various purposes

Metal	Background	Agricultural	Residential	Industrial
As	5	20	30	50
Ba	200	750	500	2000
Be	4	4	4	8
Cd	0.5	3	5	20
Cr^{6+}	2.5	8	8	*
Co	10	40	50	300
Cu	30	150	100	500
CN (free)	0.25	0.5	10	100
CN (total)	2.5	5	50	500
Pb	25	375	500	1000
Hg	0.1	0.8	2	10
Mo	2	5	10	40
Ni	20	150	100	500
Se	1	2	3	10
Ag	2	20	20	40
Sn	5	5	50	300
Zn	60	600	500	1500

* Criteria not recommended.

lower concentrations (ng mL^{-1}), graphite furnace AAS has to be used. Inductively Coupled Plasma Atomic Emission Spectrometry (ICP – AES) is being increasingly used for multielement determinations. This is because 20 – 60 elements can be simultaneously determined in a cycle time of 2 – 3 min and no dilution is required (because of the long linear dynamic range).

The Gas Liquid Chromatograph (GLC) is the basic tool for the purpose of determining organic pollutants, such as pesticides, in soils (Braithwaite and Smith, 1990). Volatile hydrocarbons can be readily examined by the method, but other substances need derivatization (such as esterification). Detection of eluted substances is made either by the Flame Ionization Detector (FID), Electron Capture Detector (ECD), or Mass Spectrometer (MS).

8.7 CHEMICALLY CONTAMINATED GROUND AND ITS TREATMENT

In 1979, Holdgate defined *Pollution* as: "The introduction by man into the environment of substances or energy liable to cause hazards to human health, harm to living resources and ecological systems, damage to structures or amenities, or interference with the legitimate use of the environment." As legally interpreted now, pollution does not require proof of *actual* harm, but simply the capacity to *cause* harm. Thus, the State can sue a polluter not only for causing a particular harm, but also for the potential harm his activities are capable of causing. Environmental knowledge thus becomes a critically important input in the case of litigation about potential harm, as such harm can only be inferred on the basis of the knowledge of *known* environmental consequences of particular activities.

Land gets contaminated by anthropogenic activities, especially industrial activities. The Department of Environment of UK defined *contaminated land*, as: "...land which represents an actual or potential hazard to health or environment as a result of current or previous use". The Greater London Council defined contaminated land as : "Land may be considered contaminated when it contains sufficient quantity of toxic or otherwise harmful material to pose a threat to the health and safety of the users of the land or workers engaged in its redevelopment..." Nowadays, the industrialized countries are engaged in massive programs of cleaning up contaminated land. While most countries subscribe to the "Polluter Pays" principle, enormously complicated legal, technical, and financial problems are involved in implementing it. In a large number of cases, the Government itself was the polluter (such as radioactive contamination from the defense nuclear installations in the USA). In most cases, it is difficult to trace the original polluter(s) and even more difficult to make them pay for the cleanup.

The following is the range of contaminants commonly encountered (Attewell, 1993, pp. 16–18):

(i) Methane and other gases from existing and abandoned landfill sites, which are combustible, toxic or asphyxiants (CO_2).

(ii) Pathogens and carcinogens.

(iii) Asbestos fibers, from the demolition of old industrial premises, factories, etc.

(iv) High concentration of heavy metals, such as cadmium, mercury, and lead, which are toxic to humans, animals, and plants.

(v) Tars, phenols, benzenes, and other organic compounds from old gasworks, which not only affect human health, but can also cause problems for new constructions.

(vi) Phytotoxic elements which prevent or inhibit plant growth.

(vii) Acidic sites, which adversely affect house construction, because of the effect of acids and sulfate on concrete.

The Interdepartmental Committee on the Redevelopment of Contaminated Land (ICRCL) U.K. (1987) (as quoted by Attewell, 1993, p. 21) gave a detailed account of the contaminants which are commonly encountered, the sites on which they are likely to occur, and the hazards they can cause. The hazards, however, are not mutually exclusive. They may sometimes combine with more serious cumulative consequences. ICRCL (1987) prescribed "trigger" concentrations for various chemical substances. For instance, the "threshold" level of concentration of phenols in soils in former coal carbonization sites which are to be used for housing, is 5 mg kg^{-1} of air-dried soil. It is safe to use the site below this level of concentration. Above this level of concentration of phenols, remedial action is recommended. Above the "Action" level of 200 (mg kg^{-1} air-dried soil), remedial action is mandatory or the land-use must be changed.

It is always difficult and expensive, and sometimes even impossible, to restore the contaminated ground to its original situation. Cleanup of contaminated ground is generally undertaken to reduce risks to public health, or to improve the environmental quality of an area, or to increase the commercial value of the land. Numerous methods of treating the contaminated ground exist, and for a particular piece of land the method of treatment is identified according to the nature and extent of contamination, cost-benefit analysis, etc.

Some of the common methods of treating contaminated land are described below (summarized from Attewell, 1993, pp. 138-148).

(i) Entombment combined with containment: this is accomplished by covering the contaminated ground with materials such as asphalt, concrete or "clean soil". Impermeable membranes may be needed to protect the covering material from the contaminated soil below.

(ii) Perimeter barriers involving, say, a bentonite slurry trench or a plastic concrete wall, to prevent the migration of toxic leachate or gas migration from the contaminated soil.

(iii) Vacuum extraction of contaminants (such as, chlorinated hydrocarbons) through a well system, and rendering them harmless by suitable methods (say, catalytic oxidation).

(iv) Stabilization/solidification using cement, lime, polymer or resin-based systems, or asphaltic systems.

(v) Chemical methods of remediation, involving oxidation, reduction, neutralization, photolysis, etc.

(vi) Bioremediation, whereby naturally occurring or genetically modified microbes are used to break down complex organic compounds (such as PCBs) to simpler and safer compounds.

(vii) Thermal treatment: direct or indirect heating, fusion or incineration.

(viii) Steam-stripping, wherein steam is injected through pipes laid in the soil to raise the temperature of the ground and evaporate the contamination.

(ix) Physical methods such as gravity separation, particle sizing and settling, etc.

8.7.1 Bioremediation through Green Plants

Containment techniques tend to be expensive and invasive. They may provide only temporary solutions and may adversely disturb the landscape.

Bioremediation through metal-accumulating plants and crops has emerged as an inexpensive and environmentally sound alternative.

Berti and Cunningham (1994) have presented a case history of utilization of this approach. "Hazardous" waste material is defined as having TCLP (Toxicity Characteristic Leaching Procedure — US EPA, 1990) Pb critical value of 5 mg L^{-1}. To bring down the soil Pb toxicity from 30 mg Pb L^{-1} in a dump to 5 mg L^{-1} level, two approaches were attempted: (i) use of lead accumulator plants, such as common ragweed (*Ambrosia artemisiifolia*), hemp dogbane (*Apocynum cannabinum*), musk or nodding thistle (*Carduus nutans*), and Asiatic dayflower (*Commelina communis*): these exhibited shoot concentrations of 400 – 1,250 mg Pb kg^{-1}; and (ii) use of soil amendments, such as lime, fertilizers, biosolids, industrial byproducts, to promote plant growth, enhance the intake of metals by plants, prevent migration of metals, reduce soil erosion and downward flow of soil water. Efforts are being made to develop more efficient soil remediation methodologies by breeding or bioengineering plants which have the ability to absorb, translocate, and tolerate Pb while producing sufficient biomass.

8.8 AMELIORATION OF LAND AFTER MINING

The mining industry is the largest producer of solid wastes; about 40,000 mines in the world process an aggregate volume of 33 billion cubic meters of rock per year. Mining affects the landscape and may cause landslides, subsidence, pollution of water and soil, lowering of groundwater, etc. Dumping of overburden, disposal of tailings, erosion brought about by rain and wind have an adverse impact on the biological productivity of the area.

Though the mining companies are required by law to submit plans and commit funds for the rehabilitation of the mined land, enforcement has not always been strict enough. It is particularly difficult in the case of artisanal miners ("here today, gone tomorrow").

Amelioration methods can be custom-made for a given situation, as follows (Chadwick et al., 1987).

Low pH (usually < 5): Amelioration by liming. Acid-tolerant species may be planted; High pH (usually > 8): Salt content may be removed by leaching. Salt/alkali-tolerant plants may be grown; Low nutrient status: Nitrogen deficiency may be ameliorated by nitrogenous fertilization or

CHAPTER

9

Soil Geochemistry in Relation to Health and Disease

*All things are poison, and nothing is without poison, and the **dosis**
alone decides that a thing is not poison*
— Paracelsus of Hohenheim (1493-1541).

9.1 ESSENTIAL ELEMENTS IN THE SOILS

Though trace elements are present in very small quantities, and *in toto*
comprise only 0.1% of animal matter, they play a crucial role in the
health and disease of humans and animals. Essential elements are trace
elements which have importance in biological processes. Inadequate intake
of essential elements leads to the impairment of the relevant physiological
functions. Supplementation of the deficient element will prevent or
ameliorate the impairment. The physiological function and the
Recommended Dietary Intake of Essential Elements is given in Table 9.1.

The abundance of the Essential Elements in the earth's crust and plant
matter are given in Table 9.2 (data regarding abundances of essential
elements in cereal grains of wheat, rye, rice, etc. is drawn from Kabata-
Pendias and Pendias, 1992). The wide difference in the degree of partitioning of trace elements between the soil and cereal grains, arises from
the differences in the mobility of the elements concerned. For instance,
the ratio of concentration between the cereal grains and soil is high for a
mobile element such as molybdenum (about 0.3), and very low for the
hydrolyzate element such as chromium (about 0.0003).

The data in Table 9.2 indicate that (i) the partitioning of elements
between rocks and plants through soils and water is critically dependent
upon the ability of the element to form soluble complexes in the soil
environment; and (ii) with the exception of selenium, the essential elements
are sufficiently abundant in the crust (more so, in mafic rocks) to meet
the physiological functions of human beings. Why the rare element

Table 9.1   Recommended dietary intake of essential elements

Element	Physiological function	Recommended daily Intake (mg d^{-1})
Calcium	Enzyme activator, electrolyte	800
Phosphorus	Nucleic acid element	800
Magnesium	Enzyme activator; electrolyte	350 (males); 300 (females)
Iron	Metallo-enzyme, enzyme activator	10 (males), 18 (females)
Zinc	– same –	15
Manganese	– same –	2.5 - 5.0
Fluorine	Enzyme inhibition (glycolysis)	1.5 - 4.0
Copper	Metallo-enzyme	0.15 - 0.5
Molybdenum	– same –	2.0 - 3.0
Chromium		0.05 - 0.2
Selenium	Enzyme protein	0.05 - 0.2
Iodine	Thyroid element	0.15

Table 9.2   Abundance of essential elements in the relevant media (μg g-1)

Element	Atomic no.	Abundance in crust	Abundance in soils	Abundance in cereals (dry wt)
Calcium	20	45000	15000	
Phosphorus	15	610	800	
Magnesium	12	16400	5000	
Iron	26	35900	40 000	17 (corn)
Zinc	30	127	90	16 - 35
Manganese	25	720	1000	12 - 80
Fluorine	9	585		0.2 - 2
Copper	29	32	30	2.2 - 6.7
Molybdenum	42	1.7	1.2	0.35 - 0.92
Chromium	24	71	70	0.014 - 0.2
Selenium	34	0.05	0.4	0. 02 - 0.45
Iodine	53	0.45	0.05 - 10	0.06 - 0.1

selenium came to be an essential element for humans remains a mystery. The author offers a fanciful idea. Since the primordial Eve hailed from northern Tanzania where the rocks (alkaline basalts and carbonatites), soils, and consequently plants, have significant quantities of selenium, it is possible that humans got "hooked" to having selenium in their food!

9.2   PEDOMEDICAL CONSIDERATIONS

Scharpenseel and Becker-Heidmann (1990) examined the role of Cation Exchange Capacity (CEC), pH, and organic matter content of soils in the mobility and fixation of inorganic cations. HAC (High Activity Clay) soils with micaceous and smectite clays have large negatively charged surface areas (about 800 m^2 g^{-1}) and correspondingly high CEC over a wide range of pH. On the other hand, soils with LAC (Low Activity Clay), candite clays, have a small surface area (< 100 m^2 g^{-1}), very low CEC, and variable charge at lower pH level. In such a situation, sorption, fixation, and release of polluting cations is largely controlled by the presence of organic matter whose ZPC (Zero Point Charge) lies at pH 3 to 3.5. Anionic pollutants can be incorporated in Ultisols only when they are rich in organic matter. Anion sorption does not take place in highly negatively charged Histosols. Anionic pollutants get fixed more readily and stay longer in Andisols than in any other kind of soils. The soil factors (charge of clays, pH, presence of humic substances, etc.) that influence the sorption and release of cationic and anionic pollutants are schematically shown in Fig. 9.1.

Some soils are characterized by deficiency or excess of some trace elements of geomedical significance (Table 9.3; after Scharpenseel and Becker-Heidmann, 1990):

Table 9.3   Deficiency or excess of some trace elements in certain types of soils

Lack or excess of element	Type of soils
Lack of Se and I	Soils of young moraines, montane soils, soils far inland from the coast
Lack of Cu,Co, and Zn	Histosols, sandy spodosols, calcareous (sand) soils
Excess of Cr and Ni	Soils of ultrabasic rocks, serpentinites, ophiolites
Excess of Ni, Cr, Pb, As	Soils near mining areas
Lack of PO$_4$, SO$_4$, and other anions	Histosols, due to lack of anion sorption capacity

(See Table 1.2 for description of soils mentioned above)

9.3   BIOGOCHEMICAL PATHWAYS OF TRACE ELEMENTS TO HUMAN BEINGS

In the ultimate analysis, all essential elements that humans need are derived from the geoenvironment (rocks, soils, water, and air) (Fig. 3.3). Diet is the principal source of essential elements, about 80%. Fluoride is an exception in that the bulk of its intake is through water. Seaspray is an important source of iodine for people who live in coastal zones, which accounts for the lower rate of goiter incidence in such areas.

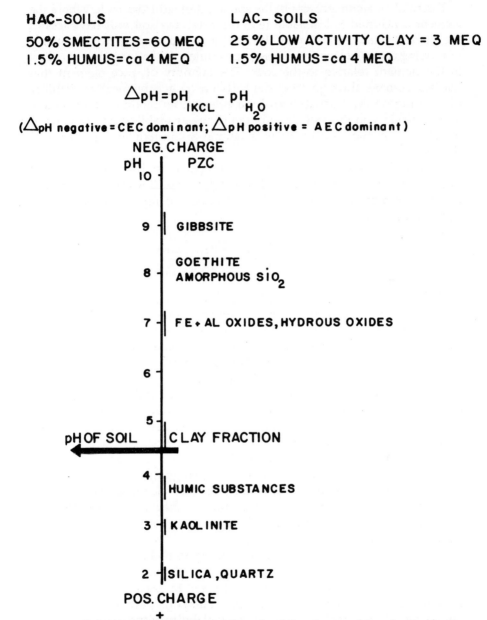

HAC–SOILS
50% SMECTITES=60 MEQ
1.5% HUMUS=ca 4 MEQ

LAC– SOILS
25% LOW ACTIVITY CLAY = 3 MEQ
1.5% HUMUS=ca 4 MEQ

$$\triangle pH = pH_{IKCL} - pH_{H_2O}$$

($\triangle pH$ negative = CEC dominant; $\triangle pH$ positive = AEC dominant)

NEG. CHARGE
pH   PZC

10

9   — GIBBSITE

8   — GOETHITE
    AMORPHOUS SiO$_2$

7   — FE+AL OXIDES, HYDROUS OXIDES

6

5

pH OF SOIL   CLAY FRACTION

4   — HUMIC SUBSTANCES

3   — KAOLINITE

2   — SILICA, QUARTZ

POS. CHARGE
+

Fig. 9.1 Schematic presentation of the role of the charge of the clays, pH, etc.
in regard to the sorption of ions
(*source:* Scharpenseel and Becker-Heidman, 1990 © CRC Press, Boca
Raton, Florida, USA).
HAC–High Activity Clays; LAC-Low Activity Clays; CEC–Cation Exchange Capacity; AEC-Anion Exchange Capacity

Essential elements present in the soil are derived from rock. Where the element concerned is leachable (e.g. fluoride), the soil will have less of the element relative to the rock. Where the element concerned is capable of forming complexes with humus (e.g. iodine), the soil is much enriched in the element relative to the rock. The quantity of trace element that reaches humans through food depends on (i) geochemical availability (depending on the leachability of the element, its mode of distribution, and availability in the rock or soil) and (ii) bioavailability (the fraction of the element in the food of plant or animal origin available to humans) (Fig. 3.4).

Biological systems have some inherent capacity to discriminate against toxic elements in their uptake. This is called the Biopurification factor. The BPF of a toxic element, X, is determined in relation to a nutrient element, Ca, as follows:

$$BPF = \frac{X_{in\ food} \times Ca_{in\ consumer}}{X_{in\ consumer} \times Ca_{in\ food}} \tag{9.1}$$

Because of the operation of BPF, the toxic element should decrease in the food chain, whereas the nutrient element should increase in the food chain relatively. Also, the meat animal serves as a good buffering agent. A toxic element present in grass will not be automatically transferred to the flesh of the grass-eating animal at the same level of concentration. Most often, the concentration is reduced.

Trace elements exist in nature in different chemical forms, and their speciation influences their toxicity (methyl mercury is more toxic than metallic mercury, and chromium as Cr (III) is an essential element needed by humans, whereas Cr (VI) is toxic to them). Besides, trace elements may undergo biotransformations in the human body and may form metallobiocomplexes. This may involve a change in the oxidation state and hence in the toxicity of the concerned trace element. Two or more elements may act synergistically, accentuating their combined effect, or antagonistically, suppressing the effect in part. For instance, zinc can moderate the toxicity of cadmium.

More and more researchers are beginning to realize that the various properties of minerals, such as size, shape, and dissolution characteristics, influence the biochemical processes which lead to disease. Hence, simply prescribing a permissible level of concentration of a particular element is not enough. The critical factor is the bioavailability of the toxic element in that medium. For instance, US EPA has prescribed 20 ppm of As in gold tailings in California as the permissible limit. It has been found that gold tailings with As concentrations as high as 3,650 ppm can still satisfy the US EPA criterion if As occurs in a form not ingestible by humans, i.e., carbonate or sulfate. On the other hand, a toxic element considered to be below the permissible threshold may indeed cause harm (e.g. cancer) if accessible in a highly bioavailable form.

9.4 TRACE ELEMENTS DEFICIENCY DISORDERS

There are several trace element deficiency disorders whose etiology and epidemiology can only be delineated through an understanding of the behavior of the concerned element in the soil. Case histories of Iodine Deficiency Disorders and Keshan disease are presented in this section.

9.4.1 Iodine Deficiency Disorders

More than a billion people, mostly living in the developing countries, are at risk from Iodine Deficiency Disorders (IDD).

Aswathanarayana (1990) gave a detailed account of the distribution of iodine in the geoenvironment and the etiology and epidemiology of the Iodine Deficiency Disorders (IDD) observed in the African context.

The concentration of iodine in the natural geoenvironment is generally low (of the order of ppb). The average concentration of iodine in different rocks is as follows: Igneous rocks — average of 40 ppb, with carbonatites having the highest concentration of 500 to 1000 ppb; metamorphic rocks — less than 10 ppb, with granulite grade metamorphic rocks having the lowest concentrations among all rocks. Iodine content of sedimentary rocks is directly dependent upon the abundance of organic matter: carbonaceous shales may contain about 10 000 ppb (10 ppm) of iodine, whereas sandstones and quartzites which contain little organic matter, have about 50 ppb of iodine.

The behavior of iodine in the geoenvironment is strongly influenced by its enrichment severalfold in soils relative to rocks. The iodine content of soils depends upon the humus content, pH, and contribution from atmospheric precipitation. Iodine in the diet is directly related to the iodine content of soils. When soil erosion takes place either because of water erosion in a hilly country or because the vegetation cover has been removed by people in search of fuelwood for cooking, the soil humus and clay component of the soil with which iodine is associated, also get washed away. Thus in areas subject to severe erosion, as in the Himalayan or Andean foothills, the soils have lost whatever little iodine they may have initially contained. Sea water contains 0.06 mg L^{-1} of iodine. Hence, sea spray, seafood and kelp are rich in iodine. Sea salt does not, however, inherit all the iodine contained in the sea water because of the volatility of iodine.

The deficiency of iodine in the geoenvironment and the consequent endemicity of IDD are common in the highland areas away from the seacoast, characterized by severe erosion. People living in the coastal areas and consuming seafood rarely suffer from IDD.

Iodine Deficiency Disorders arise when the iodine intake falls below the recommended level of 150 μg of iodine per capita per day. Goiter (enlargement of the thyroid gland) is the most well-known form of IDD. The thyroid gland produces thyroxine, which controls the growth and activity of the body. Goiter may develop not only as a consequence of iodine deficiency (the usual case), but also due to the consumption of goiterogenic foods, such as cassava, which cause goiter by impairment of the utilization of iodine by the human body. Excess calcium in the drinking water can precipitate iodine, and thus make it biochemically unavailable to man. Similarly, chemical or bacterial pollution or protein-energy undernutrition may aggravate goiter, though they may not cause it.

A dreadful form of IDD is cretinism. A cretin is a physically and mentally retarded, deaf-mute human being. Children born of parents residing in goiter-prone areas may be cretins. The results of fetal hypothyroidism include abortions, still-births, low birth weight, and congenital abnormalities such as spastic diplegia. On the basis of his studies about the incidence of goiter in Tanzania, Kavishe (1985) estimates that if the iodine intake of women in a population falls below 25 μg d^{-1}, up to 10% of the children born may develop irreversible cretinism.

The World Health Organization (WHO) graded the severity of goiter as 0, 1a, 1b, 2, and 3. Below grade 0, the thyroid is not palpable or if palpable, not larger than normal. At the other end of the scale, grade 3 corresponds to a very large goiter which is visible from a distance.

China has 425 million people who are at risk with regard to IDD. They constitute 40% of the world's population in that category. About 7 million people have serious goiter problems. There are 200,000 cretins and 8 million subcretins. The province of Sichuan is the worst affected in the country. It is now believed that there is a drop of 10 – 15 IQ points in the case of children with subcritical IDD. Since 6 million babies are born annually in the areas of endemic IDD in China, it follows that the country suffers an annual loss of at least 60 million IQ points. About 80% of the mentally disabled in China are believed to be suffering from iodine deficiency.

The case history of the village of Wangjiashan in central Hunan province in China illustrates the magnitude of human tragedy that could arise from IDD. Of the 500 people in the village, 58 are seriously retarded and 270 slightly retarded. The local people have become incapable of performing the most elementary farming jobs, and three-quarters of the children fail in primary school, so much so that the village is being run almost wholly by outsiders.

China has embarked upon a massive program of iodization of salt in order to eliminate IDD by the year 2000.

It has been estimated that more than 200 million people in Africa are at risk with respect to IDD. About 8 million people are affected by severe IDD, including 2 million cretins and 6 million mentally retarded.

Hetzel (1987) graded the prevalence of goiter and indicated the relative ameliorative measures:

(i) *Severe IDD:* Goiter prevalence of > 30%, prevalence of endemic cretinism of 1 to 10%, median level of iodine in urine < 25 μg g^{-1} of creatinine. Requires the use of iodinated oil either orally or through injection (at a cost USD 0.09 cents per capita per annum).

(ii) *Moderate IDD:* Goiter prevalence up to 30%, some hypothyroidism; mean urine levels in the range of 25 to 50 μg g^{-1} of creatinine; controllable with iodinated salt (25 to 40 mg kg^{-1} of salt) or iodized oil.

(iii) *Goiter prevalence* in the range of 5 to 20% (school children); median iodine levels in urine > 50 μg g^{-1} of creatinine; controllable with iodinated salt at concentration levels of 10 to 25 mg kg^{-1} of salt (at a cost USD 0.05 cents per capita per annum).

It should be emphasized that goiter is entirely preventable and, that too, at a small expense.

The areas characterized by severe IDD differ widely in lithology but the common factors are high elevations, severe erosion (accentuated by deforestation), and consequent low content of iodine in soils, waters, and food grown in them. The only exception is the Morogoro area in Tanzania, which is neither a highland area nor is it far from the coast. The prevalence of goiter and cretinism in this area may be traced to the extremely low content of iodine in the granulite grade metamorphic rocks. An increased incidence of the carcinoma thyroid in areas of endemic goiter is a distinct possibility in the case of Tanzania and needs to be investigated.

9.4.2 Selenium Deficiency Disorders

The geochemistry of selenium in soils has been summarized by Fergusson (1990, pp. 366-367). Selenium is often found associated with sulfur in volcanic rocks. The speciation of selenium in soils is as follows:

Acid soils, Se (-II) : selenide (HSe$^-$), and Se^{2-}; Se (0): Se
Alkaline soils, Se (IV) : selenite HSeO$_3^-$, SeO$_3^{2-}$
Se (VI) : selenate (SeO$_4^{2-}$)

The accessibility of selenate increases with higher pH. The selenite species are sorbed on iron and manganese oxides, as they form highly insoluble compounds with iron, such as Fe$_2$ (SeO$_3$)$_3$ and Fe$_2$ (OH)$_4$ SeO$_3$, whereas with selenate in alkaline soils, no such insoluble compounds are formed. Hence selenium in acid soils is immobile and unavailable to plants, whereas in the alkaline soils, selenium is more mobile and available to plants. Selenium undergoes biomethylation in soils, and the process is enhanced by the addition of organic matter. Thus, rice grown in acid soils tends to have low selenium content, and the consumption of such rice could lead to selenium deficiency disorders, such as Keshan disease in China.

Selenium protects against oxygen radical-induced damage to cellular structures. Keshan disease, a cardiomyopathic condition particularly affecting children and women of child-bearing age, has been attributed to selenium deficiency. Supplementation with sodium selenite has been found to reduce the incidence, morbidity, and fatality of the disease. On the basis of the observation of low risk for cardiovascular disease and cancer morbidity and mortality in the case of subjects with high serum selenium concentrations, it has been suggested that selenium probably gives protection against cardiovascular diseases and cancer.

9.5 TRACE ELEMENT EXCESS DISORDERS

Arseniasis arising from excess arsenic contamination in water supplies and agricultural soils, has emerged as an environmental issue of global concern (Thornton, 1998). The main "hot spots" are in the Indian subcontinent — Bangladesh where 71 million people in 41 districts are at risk, and the West Bengal province of India where 4 million people in 7 districts are affected. Other countries affected by severe arsenic toxicity are: Taiwan, Mexico, Chile, Argentina, China, etc. In the case of West Bengal and Bangladesh, arseniasis is caused by the drinking of tubewell water with high arsenic content ($60 - 3700$ μg L^{-1}) as against the Maximum Contaminant Level of 50 μg L^{-1} (Mandal et al., 1997). Arsenic-induced skin lesions, bronchitis, noncirrhotic portal fibrosis, and polyneuropathy are prevalent in the endemic areas. Cancers of the skin, lung, liver and bladder, and mental retardation in children in some countries, have been attributed to chronic arsenic exposure. Toxicity of arsenic is strongly speciation-specific; As (III) is far more toxic than As (V).

The source of arsenic may be natural: estuarine sediments (?) as in West Bengal and Bangladesh, Quaternary volcanism as in Chile, black shale as in Taiwan, and hydrothermal deposition as in the western USA. It may also be anthropogenic: coal burning as in China, and mining and smelting as in NWT, Canada.

Estuarine sediments (?) are the most likely sources of arsenic in West Bengal and Bangladesh. Estuaries act as sinks for heavy metals such as As, coming from rivers and the atmosphere. Though the normal range of As in the estuarine sediments is not high ($5 - 12$ μg g^{-1}), instances are known of the existence of very high As contents (up to 900 $\mu g\ g^{-1}$) in some estuaries. As (V) is more strongly absorbed onto sediments, but interconversion between As(V) and As (III) does take place in the sediments, depending upon the Eh. High As levels are associated with high iron levels in the sediments. As(III) is the most mobile of all As species in water. The absorption of As in the human body is high for anionic and soluble species, and low for insoluble species. Arsenic in the form of arsenate, AsO_4^{3-}, is strongly sorbed onto clays, iron and manganese oxides/hydroxides and organic matter. Flooded soils like those in West Bengal, contain arsenite salts with As (III), which is 5-10 times more soluble and hence more mobile than the corresponding arsenates. *It is also more toxic.* Thus, the kind of subsoil environment that prevails in the affected area, facilitates mobilization and entry into the groundwater of the more toxic As (III) (Aswathanarayana, 1998). Authigenic pyrite (FeS_2) is probably the most important host mineral for inorganic arsenic in estuarine sediments. The mobilization of arsenic from pyrites and its entry into the groundwater depends upon the mode of association of As with pyrites. For instance, arsenic existing as intergranular films or "paint" on pyrite, gets mobilized more readily than (say) As occupying a lattice site.

The following mitigation measures are suggested on the basis of the above considerations. A tubewell which taps an aquifer with high Eh characteristics, would contain lesser quantities of the more toxic As (III), and therefore should yield safer water. The arsenic content of water could be brought down to 30 μg AsL^{-1} by treating the water with an oxidant (say, Cl_2) and a coagulant (say, $FeCl_3$). Microbial remediation of community water supplies through the use of bacterial cultures of *Thiobacillus acidophilus* is an affordable alternative. Therapy with d-penicillamine and DMSA did not prove effective. The ability of chelation to avert the clinical effects of arsenic intoxication is yet to be established beyond doubt. On the other hand, oral treatment with retinoids and the use of selenium as antioxidant nutrient, hold promise to mitigate cutaneous arsenicism and other forms of As toxicity (Kossnett, 1998).

9.6 TOXICITY FROM CONTAMINATED SOIL

According to ICRCL (1987), the risks from chemicals in contaminated land are as follows:

- the direct ingestion of contaminated soil (mainly by children and grazing livestock), e.g. CN, As, Pb, PAHs.
- inhalation of dusts, toxic gases, and vapors from the contaminated soil, e.g. benzene, solvents, Hg, CO, HCN, H_2S, PH_3, asbestos

Table 9.4   Standards of assessment of soil contamination (in $\mu g\ g^{-1}$)

Category	A	B	C	TV
Inorganic pollutants				
CN (total free)	1	10	100	1
CN (total complex)	5	50	500	5
Br	20	50	300	
S	2	20	200	
Polycyclic aromatics				
PAHs (total)	1	20	200	
Naphthalene	0.1	5	50	
Anthracene	0.1	10	100	50
Benzo-[a]-pyrene	0.05	1	10	2.5
Chlorinated hydrocarbons				
CH total	0.05	1	10	
PCBs	0.05	1	10	
Chlorophenols (total)	0.01	1	10	
Pentachlorophenol	—	—	—	2
Pesticides				
Pesticides (total)	0.1	2	20	
Aromatic compounds				
Aromatics (total)	0.1	7	70	
Benzene	0.01	0.5	5	
Toluene	0.05	3	30	
Phenols	0.02	1	1	
Other organic compounds				
Cyclohexanes	0.1	5	60	
Pyridine	0.1	2	20	
Gasoline	20	100	800	
Mineral oil	100	1000	5000	

A = " Normal " reference value.

B = Value at which it is necessary to conduct further investigations into the form and bioavailability of the pollutant,

C = Intervention value ("trigger concentration") above which the soil definitely needs cleaning up.

TV = Target values which represent the environmental goals.

- uptake by plants of contaminants hazardous to animals and people through the food chain, e.g. Cd, As, Pb, Tl, PAHs.
- contamination of drinking water supplies,e.g. phenols, CN, SO_4, soluble metals, pesticides, e.g. atrazine in groundwater and permeation of water pipes by solvents,
- skin contact, e.g. tars, phenols, asbestos, radionuclides, PAHs, PCBs, and PCDDs,

Pytotoxicity
- SO_4^{2-}, B, Cu, Ni, Zn, herbicide residues

Fire and explosion
- CH_4 and high calorific wastes from landfills
- coal dust
- petroleum, solvents

Deterioration of building materials and services
- SO_4^{2-}, SO_3^{2-}, Cl^-, coal tar, phenols, mineral oils, solvents.

Guide values (in $\mu g\ g^{-1}$) and quality standards used in the Netherlands for assessing soil contamination by organic and inorganic substances, are given in Table 9.4.

Target values are given for the "standard soil" with 10% organic matter and 25% clay, but can be adjusted for other situations.

Economic Minerals in Soils

I must be good for something. God makes no junk
—An American schoolboy poster.

10.1   GEOCHEMISTRY OF LATERITISATION

Laterites and bauxites are products of deep tropical weathering. They are composed principally of oxides and hydroxides of Al and Fe, and are low in alkalis, alkaline earths and silica.

In 1807, Buchanan gave the name, "laterite" (meaning "brick-rock") to the earthy, vesicular and ferruginous crusts occurring in the coastal areas of Kerala province in southwest India. The term laterite is used to describe the iron-rich varieties (with iron oxides ranging from 30 to 80%) of weathering products, composed largely of hematite and goethite. Laterite is soft when moist and so can be cut into blocks. The blocks become irreversibly hard on drying and can be used as bricks. That this particular property of laterite was known to the ancients is evident from the fact that dressed laterite has been found in funerary structures in Kerala in megalithic time (about 700 B.C.). Laterite was used in the construction of the vast complex of Hindu temples of Angkor Vat in Cambodia and the Buddhist *stupa* in Borobodur in Java, Indonesia, in the tenth and eleventh centuries.

Soil scientists have introduced the term "plinthite" for the soft, iron-rich, humus-poor material which becomes hard on drying.

Laterites and bauxites have emerged as important sources of a variety of economic minerals. Bauxites are the most important source of aluminum. Ores of iron, manganese, nickel, chromium, titanium, gold, etc. are obtained from laterites.

Formation of laterites and bauxites is facilitated by a combination of "high rainfall, high temperature, intense leaching, strongly oxidizing environment, subdued topography, long duration of weathering, and

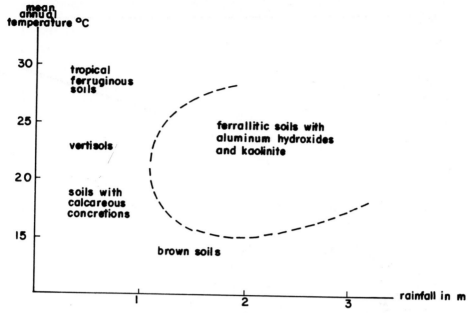

Fig. 10.1 Conditions of rainfall and temperature favourable for the development of ferrallitic soils
(*source:* Segalen, 1965; © Orstom Cahiers Pedologie)

chemically unstable rock" (Levinson, 1974). Figure 10.1 shows that ferrallitic soils with aluminum hydroxides and kaolin (with which bauxites and laterites are associated) develop under conditions of high rainfall (greater than 1000 mm) and warm temperatures (higher than 20°C).

The relationship between the mean annual rainfall (in inches) versus the weight percentage of montmorillonite, kaolinite and gibbsite (Fig. 10.2) indicates that (i) the greater the rainfall, the greater the weight percentage of bauxite in the soil (probably caused by greater desilication) and (ii) the presence of montmorillonite is a contraindication for bauxite.

Lateritization involves depletion of alkalis, alkaline earths and silica, and enrichment of Fe, Al, and Ti, and trace elements such as, Ni, Ga, Cr, Nb, etc. Economical concentrations of nickel (as in New Caledonia islands in the Pacific Ocean), and manganese (lateritoid Mn-ores of Goa, India) may have come into existence by this process.

Any model for the formation of laterites and bauxites has to account for the following: (i) removal of silica — unless silica is removed, neither laterite nor bauxite can form; (ii) separation of Al from Fe, so that aluminous (bauxitic)/ferruginous (lateritic) varieties can come into existence; and (iii) formation of laterite/bauxite profiles.

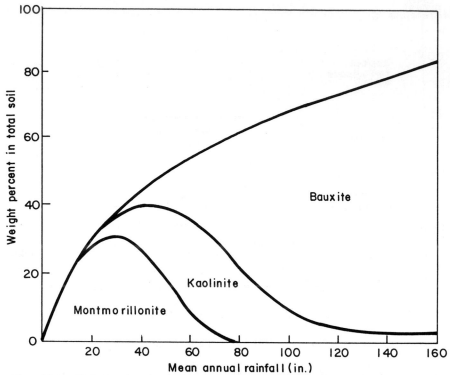

Fig. 10.2   Relationship between the mean annual rainfall versus the weight
percent of montmorillonite, kaolinite and bauxite in the soil
(*source:* Berner, 1971; © McGraw-Hill).

Removal of silica

In water, silica may occur in the form of colloidal silica or dissolved molecules of Si $(OH)_4$ or as ions. Experimental studies show that the solubility of $Si(OH)_4$ is strongly dependent on temperature (for instance, solubility increases fourfold from 0°C to 73°C) and pH (at 22°C, the solubility of Si $(OH)_4$ rises sharply after pH 9). Increased dilution at constant Al/Si ratio results in the poor adsorption of silica. Thus, alkalinity, warm temperatures, and dilute solutions favor the *transport and migration of silica.*

Separation of silica and alumina

Under conditions of acid pH (less than 4), alumina is highly soluble, whereas silica is only sparingly so. This mechanism cannot be invoked, however, for bringing about the separation of silica and alumina, as such low pH is uncommon in nature (except under conditions of acid rain). In the pH range of 5 to 9, which is more likely, alumina is virtually insoluble (and could remain behind), whereas the solubility of silica increases steadily. This is a likely mechanism to bring about the separation of SiO_2 and Al_2O_3.

Separation of iron and aluminum

The solubility of Al is higher than that of Fe under conditions of the same level of organic acid content (say, 16 ppm C) and pH (7 to 9) (vide Fig. 1.2). This mechanism can effect the separation of Fe and Al, and result in the formation of Fe-enriched laterites and Al-enriched bauxites.

While the geochemical processes described above may have contributed to the formation of bauxite, one has yet to account for different types and modes of occurrence of laterites and bauxites. Some laterites have been formed *in situ*, probably with the addition of some iron from laterally moving groundwater (Primary, autochthonous laterites). They may also be derived from primary deposits by denudation and recementation (Secondary laterites). Some occur at high elevations (Peneplain laterites), while some others occur at low level sites (Pediplain laterites). Autochthonous bauxites on igneous, metamorphic and sedimentary rocks may be primary or they may be fossil, polygenetic, altered bauxites. There are also autochthonous karst bauxites in limestone terrains.

10.2 BAUXITES

Berthier gave the name "bauxite" to the aluminous sediments in the Les Baux area in France. The term Bauxite is now used to designate the Al-rich varieties of weathering products, composed largely of gibbsite [Al $(OH)_3$], boehmite (γ – AlOOH) and diaspore (α – AlOOH). Bauxite ore should preferably contain not less than 45% Al_2O_3, and not more than 20% Fe_2O_3, and 3 – 5% combined silica.

Valeton (1972) gave a trilinear classification of bauxites in terms of the relative proportions of Al-minerals (bauxite), Fe-minerals (iron ore), and clay minerals (clay).

The layer silicates in bauxite deposits may have originated in the following ways (Valeton, 1972):

(i)   They may be of relict origin: two-layer minerals of the kaolinite group are stable, and hence can persist in the laterite environment,

(ii) They may have been caused by the desilication of clay minerals, according to the sequence, halloysite — alumina gel — gibbsite,

(iii) They have been formed due to resilicification, caused by the reaction of the silica solution with amorphous Al-hydroxides.

Trace element composition of bauxite is partly determined by the chemical composition of the parent rock. The enrichment or depletion is expressed in relation to Al. In the case of bauxites over nepheline syenites, Cr, Cu, Ga, Nb and Mo are more enriched than Al; Zr, Ti, Sc, V, Be, Mn,

Y, and Pb are less enriched than Al; Sr, La, Ba, Ca, and Mg are depleted. In the case of bauxite and kaolinite derived from the parent rock, andesite, Cr is more enriched than Al; Zr, Ga, Ti, Fe, and V are less enriched than Al, and there is depletion of Ni, Co, P, Mn, and Sr. Cr/Ni ratios in bauxite have been used to identify the nature of the source rocks in the case of karst bauxites.

Though lateritic bauxites range in age from Upper Proterozoic to Quaternary, they reached the peak of formation in the Lower Tertiary period. The temporal and spatial distribution of lateritic bauxites have been explained in terms of plate tectonic models (Valeton, 1983; Valeton et al., 1983). The high rate of sea-floor spreading in the Lower Tertiary was accompanied by inter-related phenomena such as high heat flow, high volcanic activity, and enhanced supply of CO_2 to the atmosphere (resulting in the greenhouse effect), worldwide low-relief situation, and extensive marine transgressions. This resulted in the coverage of about one-third of the continental areas by the oceans. The greater surface of the oceans led to higher evaporation and humidity.

The resultant warm, humid climate, under the influence of trade winds, promoted the formation of ferralsols (ferrallitic and allitic soils) along the coastal zones of the Deccan Peninsula (India), Guiana (S. America), Gulf coastal plain (USA), Australia and Equatorial Africa, etc. As humidity decreased inland, so did the formation of ferralsols. With increasing distance from the coast, the groundwater laterites of the coast pass laterally into ferrallitic bauxites on the hill slopes (e.g. Galiconda, Andhra Pradesh province, India).

Two types of oxisols with well-defined modes of genesis, geomorphic setting, mineralogical and other characteristics, have been recognized (Table 10.1; source: Valeton, 1983; Fig. 10.3).

Valeton (1983; Valeton et al., 1983) recognized three main types of alteration blankets:

(i) Formation of bauxites at various levels above the water table, without separation of Al and Fe; (ii) strong separation of Fe and Al in the B_{ox}–horizon, with saprolite being kaolinitic or bentonitic below the kaolinite; and (iii) total extraction of Fe and formation of flint clay below the water table.

Figure 10.4 (source: Valeton, 1983; Valeton et al., 1983) shows the relationship between the geomorphic setting, groundwater conditions, direction of drainage, and the mineralogy and geochemistry of alteration blankets.

Different types of soils may be formed on the same surface at different times. This is known as polygenetic soil formation. Valeton (1983) gave a model for possible overprint situations in respect of pre-Tertiary ferrallitic

Table 10.1 Comparison of the two types of oxisol bauxites

Description	Allitic latosols	Ferrallitic latosols
1. Topography	Coastal plains	Slightly hilly landscape
2. Genesis	"Groundwater" laterite produced by oscillating groundwater table; separation of Fe under reducing conditions; Al and Fe can become enriched in distinct zones of B_{ox}– horizon, with gel-like structures, spongy, nodular, pisolitic, etc.	Desilication by descending circulation; preservation of relict textures producing a high porosity; Fe and Al go hand in hand
3. Profiles	Relatively less thick profiles (about 10 m)	Very thick profiles (x 10 m thick)
4. Zonation	Extremely well developed, vertical and lateral zonation	Only weak vertical zonation; no lateral zonation
5. Soil horizons	Pronounced B_t - horizon; B_{ox} - horizon characterized by separation of Fe (ferricrete) and Al (alucrete)	No pronounced B_t - horizon (saprolite), and no reduction zone in B_{ox} - horizon
6. Mineralogy	Al-minerals in B_{ox} - horizon are gibbsite, followed by boehmite and diaspore; Fe minerals in B_{ox} - horizon are hematite, followed by goethite and lepidocrocite	Gibbsite is the principal Al mineral, and goethite the principal Fe mineral
7. Examples	Gujarat (western India), Surinam, Arkansas (USA), Australia, Equatorial Africa, etc.	Galiconda and Shevroy Hills (South India), bauxites on the alkaline syenites of S.E. Brazil, etc.

soils, overprinted by Lower Tertiary ferrallitic soils, and Lower Tertiary ferrallitic soils overprinted by younger soils (Fig. 10. 5).

10.3 LATERITIC ORES

10.3.1 Lateritic Nickel Deposits

Lateritic nickel deposits are associated with ophiolitic ultramafic rocks (such as peridotites, dunites, and serpentinized peridotites) emplaced in convergent margins and uplifted (thus exposing them to weathering).

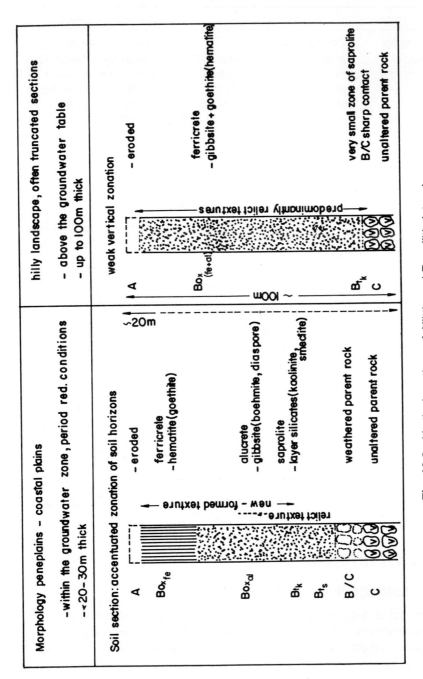

Fig. 10.3   Vertical sections of Allittic and Ferrallitic latosols
(*source:* Valeton et al., 1983)

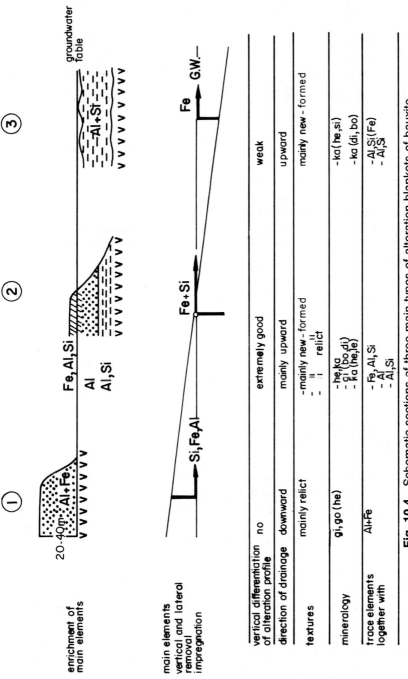

Fig. 10.4 Schematic sections of three main types of alteration blankets of bauxite
(*source:* Valeton, 1983; Valeton et al, 1983).
gi - Gibbsite; go - Goethite: he - Haematite: ka - Kaolinite; bo - Boehmite:
di - Dispore; si - Silica.

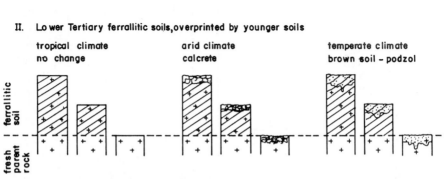

Fig. 10.5   Polyphase soil formation: a model for possible overprint situations
(*source:* Valeton, 1983; Valeton et al., 1983).

Prolonged and intense chemical weathering in a warm, humid climate, and plateau topography which permits stable, slow surface water movement and which is characterized by low rates of physical erosion, resulted in the formation of red-brown, pisolitic soils, and silica-rich boxworks (Golightly, 1981; Cox and Singer, 1986, Mineral Deposit Model 38 A). Higher grades are associated with silicate ores of nickel.

Tropical weathering of ultramafic rocks leads to the release of nickel contained in olivine and other ferromagnesian minerals (Ni^{2+} substituting for Mg^{2+} and Fe^{2+}). Nickel and cobalt thus released are transported from the upper to the lower part of the weathering crust. The infiltrating solutions become more alkaline with depth, leading to the precipitation of secondary, Ni-containing minerals, such as garnierite and redvinskite.

The following zones could be observed in the weathering profile, top-down:

Red, yellow, and brown limonitic soils (upper limonite zone may contain 0.5 – 2% Ni in iron-oxides)

Saprolites — Soft saprolite
            Hard saprolite (lower saprolite and boxwork zone contain
            2 – 4% Ni in hydrous silicates)
            Saprolitized peridotite
Fresh peridotite

The oxide and silicate ores are end members and most mineralization contains both kinds of ore minerals.

Lateritic nickel deposits occur in New Caledonia (French-administered island in South Pacific), Cuba, Indonesia, Brazil, Russia (Sakharinsk), Australia, Greece, Albania, Burundi, etc. Since nickel is toxic to plants (and animals), the presence of serpentinites and nickeliferous laterite is indicated by the sparseness of vegetation in general, besides the presence of serpentine flora which can tolerate nickel toxicity.

Large (about 40 million tons), high-grade (av. 1.4% Ni) deposits of nickel stretch like a "thin rug" on the Caledonian relief. They are derived from the Oligocene, ophiolitic ultramafic rocks, such as harzburgites (ol — cpx), dunites (ol), and lherzolites (ol-cpx-opx). The original olivine in these rocks was highly magnesian (Fo_{90}), with about 0.3% Ni, while the orthopyroxene (En_{90}) contained 0.03 – 0.06% Ni. The ophiolites were serpentinized in complex ways — part of the serpentinization may have occurred on the sea floor, but most of it is attributed to tectonism (flat faults).

Nickel got concentrated 10-30-fold in the process of lateritization of serpentine or peridotite. Generally, nickel concentrations lie at depths varying from 0.5 to 6 m below the surface. Fig. 10. 6 (source: Troly et al, 1979) shows the behavior of Cr_2O_3, Al_2O_3, and Fe_2O_3 during lateritization. It may be noted that the highest concentrations of nickel occur at the base of the weathered zone.

Garnierite [$(Mg >> Fe, Ni)_3 Si_2O_5 (OH)_4$], which is the principal ore mineral, occurs in different forms, as indicated below:

(i)   Chocolate ore: brown-colored ore (noumeite), in which Ni replaces Mg,

(ii) Green ore (nepouite), in which Ni replaces Fe in the garnierite structure.

(iii) Pimolite, which is Ni-bearing montmorillonite clay.

Up to 10% cobalt is present in the form of a cobaltiferous wad. Talc and sepiolite are the common gangue minerals, with amorphous silica occurring in the fractures cutting the serpentine masses.

The Ni content of laterite is not directly proportional to the Ni content of the ultramafic rocks and serpentinite. The occurrence of nickel deposits is strongly controlled by topography and weathering conditions. Slopes of less than 20°, and plateau topography, promote the formation of Ni deposits. Figure 10.7 indicates how the slope and the water table control the formation of Ni deposits.

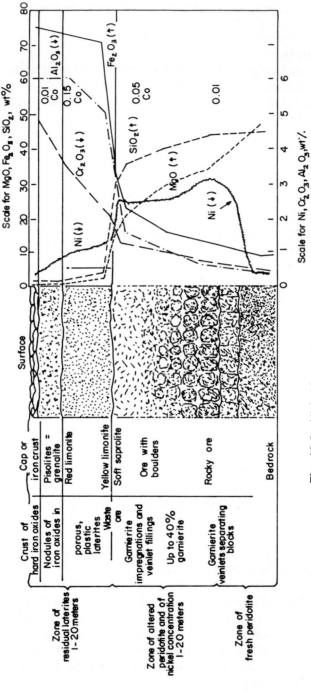

Fig. 10.6 Nickeliferous laterite profile of New Caledonia (*source*: Troly et al., 1979). Note that the highest concentrations of Ni occur at the base of the weathered zone.

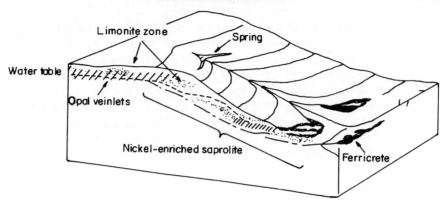

Fig. 10.7 Block diagram of laterite topography, New Caledonia. Note the control of slope and water table on the formation of nickel deposits (*source:* Golightly, 1979)

10.3.2 Lateritic Gold Deposits

Gold-bearing laterites occur in the greenstone belts. Humic acids generated in warm, humid environments mobilize and transport gold through the formation of colloidal complexes. It is not an accident that nuggets develop in the secondary environment only. Lateritic deposits are generally finegrained (< 10 μm) and low grade (about 2 g t^{-1}), e.g. Bodington deposit in Australia which annually produces more than 5 t of gold. In Tanzania, gold tends to be enriched in the mottled zone which occurs 3 – 5 m below the surface layer of red loam ("murram").

10.3.3 Lateritoid Mn and Cr ores

Lateritoid Mn ores contain minerals such as psilomelane (MnO. MnO$_2$. n H$_2$O, with 40 – 60% Mn), and amorphous wad. Goa (India) has economic deposits of this type.

Residual lateritic soils derived from chromite-bearing, ultramafic rocks under tropical and semitropical conditions, may contain up to 50% Fe, 2 – 4% Cr$_2$O$_3$, and about 0.2% Co. Eluvial and diluvial placers of chromite are known to occur in the Urals (Russia), Philippines, Zimbabwe, Albania, etc.

10.3.4 Ores in Soil Resistates

As explained earlier (Chap. 3.1), the ionic potential (Z/r) of an element determines its behavior in soils and sediments. Broadly speaking, hydrolyzate elements tend to get enriched in the soil in the form of resistate minerals.

The economic significance of the formation of soil resistates can be understood by a comparison of the behavior of U and Th. Uranium and thorium go hand in hand throughout the magmatic cycle because of the close similarity in their charges and ionic radii (U^{4+}: 1.05 Å; Th^{4+}: 1.10 Å). In the oxidizing secondary environments, they part ways. U^{4+} gets oxidized into highly mobile uranyl ion ($U^{6+}O_2)^{2+}$ and moves away, whereas Th^{4+} gets enriched in the soil in the form of resistate minerals. Thus, the C_1 zone in the weathering profile tends to have a minimum content of uranium and a maximum content of thorium. This explains why thorianite [$(Th,U)O_2$] typically occurs in resistates, whereas uraninite (UO_2) occurs as a vein mineral but never occurs in resistates.

Some of the economic minerals which occur in the soil resistates and related placers are described below:

(i) Cassiterite (SnO_2, with 78.6% Sn) is stable under exogenous conditions. Tin placers are derived from the disintegration of tin-bearing pegmatites, granites, greisen, and hydrothermal deposits. Stockwork ore deposits appear to be favorable for the formation of tin placers. Eluvial placers are formed as a result of intense and prolonged chemical weathering in hot and humid climates. Weathering can penetrate to depths of 200 m. Apart from cassiterite, the eluvial placers may contain columbite, tantalite, wolframite, scheelite, gold, topaz, tourmaline, zircon, magnetite, etc. The thickness of such placers is usually about 20 – 30 m but may occasionally reach 60 – 80 m. The average content of cassiterite is 0.5 – 1.5 kg m^{-3}, though in the rich, lower levels, the concentrations may go up to 5 – 6 kg m^{-3}. Malaysia and Indonesia have large placer deposits of cassiterite.

(ii) Ilmenite ($FeTiO_3$, containing 48.6 – 57.3% TiO_2, up to 6% Fe_2O_3, and some MgO) and rutile (TiO_2) and zircon (Zr SiO_4, with 67.2% of ZrO_2) may occur as resistates in soils, though economic concentrations of these minerals tend to develop after they have been subjected to some kind of stream or wave action. Heavy mineral sands containing ilmenite, monazite, zircon, garnet, etc. occur along the east coasts of Mozambique (Chinde to Quinga, for a stretch of 600 km), Australia (Sydney to Fraser Island), east Brazil (Espirito Santo province), Sri Lanka (Pulmoddai), west and east coasts of India (Chavara, Manavalakurichi, Uppada), etc.

10.4 ARTISANAL MINING OF ECONOMIC SOIL MINERALS

The contribution of small-scale or artisanal mining to the overall mining output in the world is estimated to be 10-16%. In general, the percentage contribution with respect to industrial minerals is higher than for metallic minerals. It has been estimated that about 6 million persons in the world are engaged in artisanal mining. In some countries, the contribution of artisanal mining is of considerable economic significance. For instance, Peru has about 3,000 small-scale mines which produce 100% of the antimony, 90% gold, 15% tungsten, etc. of the national production (Gocht, 1980, Natural Resources and Development, v.12, pp. 7-18). About 90% of Brazil's gold production (70 t, worth about USD 800 million) is attributable to artisanal mining.

Capital-intensive, mechanized mining is not cost effective for the exploitation of small deposits, even of high grade. Such a deposit is amenable for small-scale mining, particularly where the deposit occurs at or close to the surface, and where the mineral could be won by simple methods, such as hand-picking, panning, sluicing, etc. It therefore follows that artisanal mining may be a cost-effective option in the case of several economic minerals occurring in the soil.

Small-scale mining has several advantages: it is labor intensive, can be initiated on any scale, with simple technology, at low capital cost, low consumption of energy, and short lead time, and without expensive imported equipment. It can also promote local industries. This, however, does not mean that small-scale mining is the panacea for the problems of the developing countries. Artisanal mining suffers from the following serious disadvantages: (i) it tends to be haphazard, since in most countries there is no systematic exploration activity to support small-scale mining; (ii) destructive exploitation by the gouging of rich pockets; (iii) low recovery rates; (iv) low labor productivity (about 4% of highly mechanized mines); and (v) non-extraction of valuable byproducts which are therefore irretrievably lost to the country.

In future, vegetative methods of reclamation of mined land may emerge as a significant, economically viable, and employment-generating activity.

10.4.1 Mercury Pollution Due to Artisanal Gold Mining

The issues involved in artisanal mining can be examined with Tanzania as a case. An estimated 100,000 artisanal miners (half of whom are probably seasonal) produce an estimated 4,000 kg of gold (worth about USD 50 million) annually, using the mercury amalgam method. Though this works out to about USD 500 per capita per annum (inclusive of the cost of mercury and other expenses), this is still an attractive proposition for the miners as the GDP per capita of Tanzania is only USD 110 (1992). The amount of mercury pollution in Tanzania is estimated to be of the order of one mg m^{-2} annually (Aswathanarayana, 1995, p. 177).

Usually, 6 – 8 kg of mercury is used for amalgamation to extract one kg of gold. Most of the mercury is recovered by simple squeezing and sublimation, but about 1.5 kg of mercury is irretrievably lost to the environment per kg of gold extracted (the Brazilian average is 1.32 kg of mercury). This leads to mercury pollution in soils, sediments, waters, and biota. The bacteria (such as *Enterobacter aerogenes* and *Escherichia coli*) present in the soils, sediments, human feces, etc. convert the metallic mercury to the highly toxic form of methyl mercury. Mercury compounds may enter the human body through inhalation, ingestion of food and water, and transfer through the skin. The intake and uptake of mercury are highly species sensitive. The organomercury, particularly methyl mercury, $CH_3 Hg^+$, is more easily absorbed and is far more toxic than elemental mercury (Hg^0) and divalent inorganic mercury (Hg^{2+}). In the food chain, methyl mercury gets concentrated in fish (about 80% of mercury in fish is in the form of the more toxic methyl mercury). The FAO/WHO permissible tolerable level of mercury exposure has been set at 0.3 mg/ week, with methyl mercury constituting not more than two-thirds of it (i.e., 0.2 mg).

The critical organs affected by mercury intoxication are the lungs, kidneys, and the brain. The effects of mercury on the respiratory tract are coughing, bronchial inflammation, chest pain, and in severe cases, respiratory arrest. Methyl mercury causes the disintegration of cells within the brain, and consequently affects the sensory, visual, auditory and coordination control functions of the brain. This leads to loss of coordination in walking, slurred speech, loss of hearing, blindness, coma, etc. (Fergusson, 1990, p.542).

Laterites are wide-spread in several tropical, developing countries, and some countries like Zambia have vast dumps of copper tailings with quantities of gold too small to be extractable by the cyanidation process. Carbon-in-pulp and carbon-in-leach technologies are capable of extracting gold from such sources.

10.4.2   Innovative Technologies Suggested

It is possible to improve the efficiency of small-scale mining, while concomitantly reducing its deleterious consequences, by the adoption of the following innovative approaches:

1. Developing simple techniques of prospecting which could be used by semi-skilled labor, e.g. use of smoky quartz as indicator of cassiterite-lepidolite pegmatites, and looking for cassiterite resistate in the soils near pegmatite. Training of miners on-site about simple methods of mineral search and extraction.

   Using a portable X-ray fluorescence analyzer, it is possible to make a quick and fairly accurate on-site assay of several ore metals in the material mined or to be mined by a miner. Such an assay can serve two purposes: (i) to make the miner aware of the economic value of the material already mined by him (through a knowledge of what kind of ore metals and in what concentrations occur in the material mined by him), and (ii) to advise him as to what kind of material he should be mining in order to get greater returns.

2. Research and development to design improved methods of ore search and ore extraction relevant to small-scale mining. Placer gold is a case in point. An artisanal miner can extract gold only if it is coarse grained (say, 30 μm) and high grade (say, about 25 g m^{-3}). He uses the mercury amalgam method of extraction which is highly polluting. New carbon-in-pulp and carbon-in-leach technologies have several advantages: (i) they are capable of extracting fine-grained gold (about 10 μm) and at low concentrations (about 2 g t^{-1}); (ii) they are environmentally benign. These technologies need to be adapted for small-scale operations. In extremely dry areas, pneumatic methods of gold separation have to be developed.

3. Using mobile units for preconcentration and extraction on site: Truck-mounted, diesel-powered, self-contained, ore-dressing modules are taken to the site of the artisanal mining and the ore is concentrated/extracted on site. The mobile unit can be owned and operated by a cooperative or a private company. A part (say, one-third) of the output could be collected in kind towards service charge due to the mobile unit and the royalty due to the government. As the recovery rates by the mobile unit are at least 2-3 times higher than by manual methods, the artisanal miner is still left with considerably more output than he would have been able to achieve on his own. Such a use of innovative technology makes everybody happy — the miner is happy because he gets more money with less labor, the company owning the module is happy because they get their service charge, the Government is happy because not only does it receive royalties, it can also can keep track of mineral production, and the community is happy that there is less degradation of the environment. Such units are in operation for the recovery of antimony in the Andean mines in South America.

4. The 'Portable" gold plant developed by Libenberg, Rundle and Storey of San Martin mining company (*Mining Magazine*, July, 97, pp. 8-10) is a veritable godsend for small-scale gold miners. The salient points of the plant are as follows: San Martin's claims encompasses two dumps around Bonda, Kenya, with 250,000 t of material, grade: 1 – 3 g t^{-1}; Carbon-in-pulp/carbon-in-leach technique; Capacity of the plant: 10,000 t/month; production cost: USD 150/oz. The total steel requirement (12 t) for tanks, baffles, agitator mountings, and the pumps and piping, were brought from South Africa in one container and erected on site. Dump material is reclaimed by high-pressure water. In the first leach tank, lime (5 kg t^{-1}) and sodium cyanide (0.5 kg t^{-1}) are added. The residence time in each of the absorption tanks is approximately 1.3 h at a throughput of 10,000 t/month, and carbon concentration of 15 – 20 g L^{-1}. The

References

Abrol, I.P., D.R. Bhumbla and K.S. Dargan (1973). Reclaiming alkali soils. *Tech. Bull.* no.2. Karnal, India: Central Soil Salinity Research Inst.

Adams, F. (ed.) (1984). *Soil Acidity and Liming.* 2nd ed., Madison, WI, USA: Amer. Soc. Agron.

Alloway, B.J. (ed.) (1990). *Heavy Metals in Soils.* Glasgow: Blackie & Son.

* Alloway, B.J. and D.C.Ayres (1993). *Chemical Principles of Environmental Pollution.* London: Chapman & Hall.

Andrews-Jones, D.A. (1968). The application of geochemical techniques to mineral exploration. *Colorado School of Mines, Mineral Industry Bulletin,* **11** (6): 1-31.

* Anon. (1990). *Saline Agriculture — Salt-tolerant Plants for Developing Countries.* Washington, D.C.: Nat. Acad. Press.

Archer, A.A., G.W. Luttig, and I.I. Snezkho (eds.) (1987). *Man's Dependence on the Earth.* Nairobi-Paris: UNEP-UNESCO.

Arnolds, E. and Jansen, E. (1992). New evidence for changes in the macromycete flora of the Netherlands. *Nova Hedwigia,* **55:** 325-351.

Aswathanarayana, U. (1988). Natural radiation environment in the Minjingu phosphorite area, Northern Tanzania. In: Jul Låg (ed.), *Health Problems in Connection with Radiation from Radioactive Matter in Fertilizers, Soils and Rocks,* pp. 79-85. Oslo: Norwegian University Press.

* Aswathanarayana, U. (1995). *Geoenvironment: An Introduction.* Rotterdam: A.A. Balkema.

Aswathanarayana, U. (1998). Etiology of drinking water induced hyperkeratosis in parts of West Bengal,India: Proc. 3rd Int Conf. on Arsenic Exposure and Health Effects, San Diego, Calif., July 12-15, 98, p. 131.

* Attewell, P. (1993). *Ground pollution — Environment, Geology, Engineering and Law.* London: E & FN Spon.

---

* Suggested reading

Ayers, R.S. and D.W. Wescot (1985). *Water Quality for Agriculture.* FAO Irrigation and Drainage Paper 29, Rev.1. Rome: FAO.

Baes, C.F. III, R.D. Sharp, A. Sjoreen and R. Shor (1984). *A Review and Analysis of Parameters for Assessing Transport of Environmentally Released Radionuclides through Agriculture.* ORNL 5786. Oak Ridge, TN, USA: U.S. Dept. of Energy, Oak Ridge National Lab.

Banin, A., and J. Navrot (1975). Origin of life: Clues from relations between chemical composition of living organisms and natural environments. *Science.* **189**: 550-551.

* Barrow, Chris (1987). *Water Resources and Agricultural Development in the Tropics.* England: Longman's.

Bates, R.L. and J.A. Jackson (1982). *Our Modern Stone Age.* Kaufman, Los Altos, Calif., USA, 136 pp.

Becker-Heidman, P. and H.W. Scharpenseel (1990). Effects of radioactive radiation caused by man. In: Jul Låg (ed.) *Geomedicine,* pp. 170-184. Boca Raton, USA: CRC Press.

Beniston, M. (ed.) (1994). *Mountain Environments in Changing Climates,* London & New York: Routledge Publ. Comp., 492 pp.

Berner,R.A. (1971). *Principles of Chemical Sedimentology.* New York: McGraw-Hill.

Berti,W.R. and S.D. Cunningham (1994). Remediating soil lead with green plants. In C. Richard Cothern (ed.) *Trace Substances, Environment and Health,* pp. 43-51. Northwood, U.K.: Science Reviews.

Bradbury, N.J., and D.S. Powlson (1994). The potential impact of global environmental change on nitrogen dynamics in arable systems. In: *Soil Response to Climate Change.* M.D.A. Rounsevell and P.J. Loveland (eds.), NATO ASI Series I: Global Environmental Change, v. 23, Heidelberg, Germany: Springer-Verlag, , pp. 137-154.

Braithwaite, A. and F.J. Smith (1990). *Chromatographic Methods* (4th ed.). London: Chapman and Hall.

Brinkman, R. and N. van Breemen (1988). *Processes in Soils* (Lecture notes). The Netherlands: Agricultural University of Wageningen.

Brinson, M.M., R.R. Christian and L.K. Blum (1995). Multiple states in the sea-level induced transition from terrestrial forest to estuary. *Estuaries,* 18: 648-659.

Brummer, G.W. (1986). Heavy metal species, mobility and availability in soils. In: M.Bernhard, F.E.Brinckman and P.J. Sadler (eds.) *The Importance of Chemical Speciation in Environmental Processes,* pp. 169-192. Dahlem Konferenzem. Berlin: Springer-Verlag.

Buringh, P., H.D.J. van Heemst and G.J. Staring (1975). *Computation of Maximum Food Production in the World.* The Netherlands: Agr. Univ. of Wageningen.

Chadwick, M.J., N.H. Highton and J.P. Palmer (eds.) (1987). *Mining Projects in Developing Countries.* Stockholm: Beijer Institute.

Iodine Deficiency in Tanzania, 19. Dar es Salaam, Tanzania: TFNC-SIDA Report.

Kirkby, M.J., and R.P.C. Morgan (eds.) (1980) *Soil Erosion*. Chichester: Wiley.

Knowles, R (1993). Methane: processes of production and consumption. In: *Agricultural Ecosystem Effects on Trace Gases and Global Climate Change*, ASA Sp. Publ. 55, Madison, WI: Amer. Soc. Agronomy, pp. 145-156.

Korobova, E., A. Ermakov and V. Linnik (1998). ^{137}Cs and ^{90}Sr mobility in soils and transfer in soil-plant systems in the Novozybkov district affected by the Chernobyl accident. *App. Geochem.*, **13**: 803-814.

Kosnett, M.J. (1998). Clinical approaches to the treatment of chronic arsenic intoxication: from chelation to chemoprevention: Proc. 3rd Int Conf. on Arsenic Exposure and Health Effects, San Diego, Calif., July 12-15, 98, p. 46.

Kovda, V.A. (1980). *Land Aridization and Drought Control*. Boulder, Colo., USA: Westview Press.

* Låg, Jul (ed.) (1987). *Commercial Fertilizers and Geomedical Problems*. Oslo: Norwegian Univ. Press.

Lal, Rattan (1990). *Soil Eerosion in the Tropics*. New York: McGraw-Hill.

Lal, R. et al. (eds.) (1994) *Soils and Global Change*. Chelsea, MI: Lewis Publishers.

Le Houerou, H.N. (1986) Salt-tolerant plants of economic value in the Mediterrenean Basin. *Reclamation and Vegetation Res.*, **5**: 319-341.

Leith, H. (1973). Primary production: Terrestrial ecosystems: *Human Ecology*, **1**: 303-332.

Levinson, A.A. (1974). *Introduction to Exploration Geochemistry*. Wilmette, MICH, USA: Applied Publ. Ltd. (2nd ed. appeared in 1980).

Maas, E.V. (1988). Crop salt tolerance. In: *Agriculture Salinity Assessment and Management Manual*. Amer. Soc. Civil Engg. (ASCE)

Maas, E.V. and G.J. Hoffman (1977). Crop salt-tolerance and current assessment. *J. Irrig. Drainage Div., ASCE.* **103**(2): 115-134.

Manahan, S.E. (1991). *Environmental Chemistry* (5th ed.). Chelsea, Michigan, USA: Lewis Publishers.

Mandal, B.K. et al .(1997). Chronic arsenic toxicity in West Bengal: *Curr. Sci.* (Ind.), **72**(2): 114-117.

Mann, M., R.S. Bradley and M.K. Hughes (1998). Global scale temperature patterns and climate forcing over the past six centuries: *Nature*, **392**: 779-788.

Miura, N., M.R. Madhav and K. Koga (1994). *Lowlands: Development and Management*. Rotterdam: A.A. Balkema.

Morgan, R.P.C. (1986). *Soil Erosion and Conservation*. England: Longman's, 298 pp.

Neilson, R.P. and D. Marks (1994). A global perspective of regional vegetation and hydrologic sensitivities from climate change: *J. Vegetation Sci.*, **5**: 715-730.

Noronha, L. (1995). Ecological management in a mine in Goa, India — a case study of afforestation. *Geoscience & Development*, **2**: 13-18.

Ong, H.L., V.E. Swanson and R.E. Bisque (1970). Natural organic acids as agents of chemical weathering. *U.S. Geol. Surv. Prof. Paper 700 C*: C-130-C-137.

O'Neill, Peter (1985). *Environmental Chemistry*. London: George Allen & Unwin.

Paces, T. (1991). Anthropogenic effects on mass balance of weathering and erosion. In: O. Selinus (ed.), *Second Int. Symp. on Environmental Geochemistry*: Uppsala, Sept. 91.

Patterson,C.C. (1980). Alternative perspective — Lead pollution in the human environment: Origin, extent and significance. In: *Lead in Human Environment*, pp. 265-349, Washington, D.C.: National Academy of Sciences.

Pedro, G. (1984). La gense des argiles pedologiques, ses implications mineralogiques, physico-chimiques et hydriques. *Sci. Geol. Bull.* **37**(4): 333-347.

Rabbinge, R., and M.K. van Ittersum (1994). Tension between aggregation levels. In: *The Future of Land: Mobilizing and Integrating Knowledge for Land-use Options*. L.O. Fresco, L. Stroosnijder, J. Bouma, and H. van Keulen (eds.), Chichester, UK: John Wiley, pp. 31-40.

Rawlins, S.L. (1977). Uniform irrigation with low-head bubbler system. *Agriculture and Water Management*. **1**: 167-178.

Richards, L.A. (ed.) (1954). *Diagnosis and Improvement of Saline and Alkaline Soils* (USDA Handbook, 60). Riverside, Calif., USA: U.S. Soil Salinity Lab.

Rounsevell, M.D.A. and R.J.A. Jones (1993). A soil and agroclimatic model for estimating machinery work days: the basic model and climate sensitivity. *Soil and Tillage Res.*, **26**: 179-191.

Rounsevell, M.D.A. and P.J. Loveland (eds.) (1994) *Soil Responses to Climate Change*. NATO ASI Series 1, Global Environmental Change, v.23. Heidelberg, Germany: Springer-Verlag.

Santvoort, G.P.T.M. van (ed.) (1994). *Geotextiles and Geomembranes in Civil Engineering*. Rotterdam: A.A. Balkema.

Scharpenseel, H.W. and P. Becker-Heidman (1990). Pollution as a geomedical factor. In: Jul Låg (ed.) *Geomedicine*, pp. 141-162. Boca Raton, USA: CRC Press.

Schwab, A.P. and W.L. Lindsay (1983). Effect of redox on the solubility and availability of iron. *Soil Soc. Am. J.* **47**: 201-205.

Schwab, A.P. and Lindsay, W.L. (1983). The effect of redox on the solubility and availability of manganese in a calcareous soil. *Soil Soc. Am. J.* 47: 217-220.

Schuphan, W. (1972). *Effects of intensive fertilizer use on the human environment.* Panel Discuss., FAO-AGL: FHE/72/9, 15 pp. Rome: FAO.

Segalen, P. (1965). Les produits aluminex *dans les sole de* in zone tropicale humide: *1.2 Orstom Cahiers Pedol.* 3: 149-176 & 179-205.

Sehgal, J. and I.P. Abrol (1994). *Soil Degradation in India: Status and Impact.* Ind. Coun. Agr. Res., and Oxford-IBH, New Delhi.

Shainberg, I. and J.D. Oster (1978) *Quality of Irrigation Water.* Bet Dagan, Israel: Intl. Irrig. Info. Ctr.

Shuval, H.I. et al. (1986). *Waste Water Irrigation in Developing Countries.* Tech. Paper no. 51. Washington, D.C.: World Bank.

Sims, J.T. and S.E. Heckendorn (1991). *Methods of Soil Analysis.* Newark, DE, USA: University of Delaware Soil Testing Laboratory Cooperative Bull. # 10 (revised).

* Sposito, G. (1989). *The Chemistry of Soils.* New York: Oxford Univ. Press.

Stern, P.H. (1979). *Small Scale Irrigation: A Manual for low-cost water technology.* Nottingham, U.K.: Intermediate Technology Publ. & Russell Press.

Stevenson, F.J. (1982). *Humus Chemistry.* New York: Wiley.

* Stulz, Roland and Kiran Mukherji (1993). *Appropriate Building Materials.* St. Gallen, Switzerland: Swiss Centre for Development Cooperation in Technology and Management.

* Swaminathan, M.S. (1991). *From Stockholm to Rio de Janeiro — the Road to Sustainable Agriculture.* Monograph no. 4. Madras, India: M.S. Swaminathan Research Foundation.

Sytchev, K.I. (ed.) (1988). *Water Management and Geoenvironment.* Paris-Nairobi: UNESCO-UNEP.

Takkar, P.N. and R.L. Bansal (1987). Evaluation of rates, methods and sources of zinc application to wheat. *Acta Agron.* 36: 277-283.

Thornton, I. (1998). Arsenic in the global environment: looking towards the millenium: Proc. 3rd Int Conf. on Arsenic Exposure and Health Effects, San Diego, Calif., July 12-15, 98, p. 3.

Tinker, P.B. and J.S.I. Ingram (1994) Soils and global change: An overview. In: *Soil Response to Climate Change.* M.D.A Rounsvell and P.J. Loveland,. (eds.) NATO ASI Series I: *Global Environmental Change,* v.23, Berlin-New York: Springer-Verlag, pp. 3-12.

Topp, G.C. and J.L. Davis (1985). Time-domain reflectometry (TDR) and its application to irrigation scheduling. In: D. Hillel (ed.) *Advances in Irrigation,* v.3, Orlando, Florida, USA: Academic Press.

Troly, G., M. Esterle, B. Pelletier, and W. Reibell (1979). Nickel deposits in New Caledonia — some factors influencing their formation. Int. Laterite Symp., New Orleans, 1979, pp. 85-119.

US Environmental Protection Agency (1990). Federal Register, part V, 40 CFR part 261, **55** (126): 26986-26998.

Valeton, I. (1972). *Bauxites*. Amsterdam: Elsevier.

Valeton, I. (1983). Paleogeographical interpretation of the world-wide distribution of oxisols (laterite bauxites) in the Lower Tertiary: *Travaux ISCOBA*. **13** (18): 11-22.

Valeton, I., B. Stutze and M.R. Goldberg (1983). Geochemical and mineralogical investigations of the Lower Jurassic flint clay-bearing Mishar and Ardon formations, Mathteh Ramon, Israel. *Sed. Geol.* **35**: 105-152.

* Wild, Alan (ed.) (1988). *Russell's Soil Conditions and Plant Growth* (11th. ed.) ELBS. England: Longman's.

* Wild, Alan (1993) *Soils and the Environment*. Cambridge, U.K.: Cambridge Univ. Press, 287 pp.

Williams, J.R., P.T. Dyke, W.W. Fuchs, V.W. Benson, O.W. Rice and E.D. Taylor, (eds.) (1990). EPIC-Erosion/Productivity Impact Calculator: 2, User Manual, Tech. Bull. 1768, US Dept. of Agriculture.

Woodruff, N.P. and F.H. Siddoway (1965). A wind erosion equation. *Soil Sci. Soc. Amer. Proc.*, **29**: 602-608

Author Index